Laboratory Manual for Geotechnical Characterization of Fine-Grained Soils

This manual presents procedures for performing advanced laboratory tests on fine-grained soils. It covers characterization tests, which determine soil composition and quantify the individual components of a soil; and behavioral tests such as the Atterberg Limits tests, which demonstrate how the fines fraction of a soil reacts when mixed with water, and the Linear Shrinkage Test, which demonstrates how much a soil shrinks. The material goes beyond traditional evaluation of basic soil behavior by presenting more advanced laboratory tests to characterize soil in more detail. These tests provide detailed Compositional Characteristics which identify subtle changes in conditions and vertical variations in the soil, and which help to explain unusual behavior.

- A unique compilation of information on key soil tests
- Combines characterization tests with behavior tests

The book suits graduate students in geotechnical engineering, as well as practitioners and researchers.

Alan J. Lutenegger is Emeritus Professor of Civil and Environmental Engineering at the University of Massachusetts-Amherst, USA. He is a registered professional engineer and a Fellow of the American Society of Civil Engineers. He is author of *Soils and Geotechnology in Construction* and *In Situ Testing Methods in Geotechnical Engineering*, both published by CRC Press.

Laboratory Manual for Geotechnical Characterization of Fine-Grained Soils

Alan J. Lutenegger

CRC Press
Taylor & Francis Group
Boca Raton London New York

CRC Press is an imprint of the
Taylor & Francis Group, an **informa** business

Cover image: the author

First edition published 2023
by CRC Press
6000 Broken Sound Parkway NW, Suite 300, Boca Raton, FL 33487-2742

and by CRC Press
4 Park Square, Milton Park, Abingdon, Oxon, OX14 4RN

CRC Press is an imprint of Taylor & Francis Group, LLC

ISBN: 978-1-032-20345-4 (hbk)
ISBN: 978-1-032-20346-1 (pbk)
ISBN: 978-1-003-26328-9 (ebk)

DOI: 10.1201/9781003263289

Typeset in Sabon
by Deanta Global Publishing Services, Chennai, India

Contents

Preface   xvii
Acknowledgments   xix
Introduction   xxi

PART I
Determination of Compositional Characteristics   I

1 **Water Content**   3

Background 3
Water Content by Convection Oven 3
  Introduction 3
  Standard 5
  Equipment 5
  Procedure 5
  Calculations 5
  Example 5
  Discussion 6
Water Content by Microwave Oven 8
  Introduction 8
  Standard 8
  Equipment 9
  Procedure 9
  Calculations 9
  Example 9
  Discussion 10
Water Content by Portable Carbide Gas
  Method – "Speedy" Moisture Meter 11
  Introduction 11
  Standard 13
  Equipment 14
  Procedure 14

Calculations 15
Example 15
Discussion 15
References 17
Additional Reading 18

2 **Phase Relationships of Soils: Determining Unit Weight,
   and Calculating Voids Ratio, Porosity and Saturation**   19

Background 19
Phase Relationships in Soils 20
  Phase relationships based on mass 20
    Water Content 20
  Phase relationships based on volume 21
    Voids Ratio 21
    Porosity 21
  Phase relationships combining mass and volume 22
    Wet Density 22
    Dry Density 23
    Saturated Density 23
    Buoyant Density 23
Determining Total Unit Weight on trimmed specimens 24
  Standard 24
  Equipment 24
  Procedure 25
  Calculations 25
  Example 27
  Example 28
    Water Content 29
    Wet (Total) Unit Weight 29
    Dry Unit Weight 29
  Discussion 29
Determining Total Unit Weight using the Eley Volumeter 30

3 **Density (Unit Weight) of Solids (Specific Gravity)**   33

Introduction 33
Standard 33
Equipment 34
Preparation 34
Calibration of Pycnometer 35
Procedure 35
Calculations 36
Example 36

Discussion 37
  Unit Weight of water 38
  Unit Weight of solids 38
Reference 39

4 Grain-Size Distribution   41

Background 41
Coarse Fraction – Sieve Analysis 42
  Introduction 42
  Standard 42
  Equipment 42
  Procedure 42
  Calculations 44
  Example 45
  Discussion 46
Fine Fraction – Hydrometer Analysis 47
  Introduction 47
  Standard 48
  Equipment 48
  Procedure 48
  Calculations 50
  Example 50
  Discussion 51
Fine Fraction – Pipette Analysis 51
  Introduction 51
  Standard 52
  Equipment 52
  Procedure 52
  Calculations 54
  Example 55
  Discussion 55
References 57
Additional Reading 58

5 Organic Content   59

Background 59
Organic Content by Loss-on-Ignition (LOI) 61
  Introduction 61
  Standard 61
  Equipment 61
  Procedure 62
  Calculations 62

Example 63
Discussion 63
Organic Content by Hydrogen Peroxide 64
  Introduction 64
  Standard 65
  Equipment 65
  Procedure 65
  Calculations 66
  Example 66
  Discussion 66
References 67
Additional Reading 68

6 **Carbonate Content**   71

Background 71
Carbonate Content by Rapid Carbonate Analyzer 72
  Introduction 72
  Standard 72
  Equipment 73
  Calibration of the Carbonate Analyzer 73
  Procedure 75
  Calculations 75
  Example 76
  Discussion 76
Carbonate Content by Chittick Apparatus 76
  Introduction 76
  Equipment 76
  Setup 77
    Reservoir fluid 77
    Dilute hydrochloric acid 77
  Procedure 79
  Calculations 79
  Example 80
  Discussion 80
References 82
Additional Reading 82

7 **pH of Soils**   85

Introduction 85
Standard 85
Equipment 86
Procedure 86

Calculations 86
Discussion 87
Reference 88
Additional Reading 88

8 Cation Exchange Capacity   **91**

Introduction 91
Standard 92
Equipment 92
Reagents 92
Procedure 93
Calculations 94
Discussion 94
References 95
Additional Reading 96

9 Specific Surface Area   **97**

Introduction 97
Standard 98
Preparation 98
Equipment 98
Procedure 99
Calculations 100
Example 101
Discussion 101
    Kaolinite 104
    Montmorillonite 105
References 106
Additional Reading 108

10 Methylene Blue Test   **111**

Introduction 111
Standard 111
Equipment 111
Preparation 112
Procedure 112
Calculations 113
Example 113
Discussion 114
References 114
Additional Reading 115

11 Pore Fluid Salinity   117

Introduction 117
Pore Fluid Salinity by Refractometer 118
  Standard 118
  Equipment 118
  Preparation 120
  Procedure 120
    Part 1: Extracting pore fluid 120
    Part 2: Measurement 120
  Calculations 121
  Discussion 122
Pore Fluid Salinity by Electrical Conductivity 122
  Standard 124
  Equipment 124
  Preparation 124
    Calibration of the cell 124
    Extracting pore fluid using a soil press 124
    Extracting pore fluid using a centrifuge 125
  Procedure 125
  Calculations 125
  Discussion 126
References 127
Additional Reading 128

PART II
Determination of Behavioral Characteristics   131

12 Consistency of Fine-Grained Soils: Atterberg Liquid and
Plastic Limits   133

Background: The Atterberg Concept 133
Liquid Limit by Casagrande Drop Cup 134
  Introduction 134
  Standard 134
  Equipment 135
  Calibration of Casagrande Drop Cup 135
  Procedure 136
  Calculations 137
  Example 137
  Discussion 138
Liquid Limit by Fall Cone 139
  Introduction 139

Standard 141
Equipment 141
Procedure 142
Calculations 142
Example 143
Discussion 144
Plastic Limit by Rolling Thread 147
Introduction 147
Standard 148
Equipment 148
Procedure 148
Calculations 150
Example 150
Discussion 150
Plastic Limit by Fall Cone 150
Introduction 150
Standard 151
Equipment 151
Procedure 151
Calculations 152
Example 152
Discussion 152
One-Point Liquid Limit tests 153
Discussion of using Atterberg Limits 154
Aterberg Limits and Unified Soil Classification 155
Activity 158
Plasticity Ratio 158
Discussion 159
References 160
Additional Reading 163

13 Shrinkage Limit   **165**

Background 165
Shrinkage Limit by Mercury Method 166
Standard 166
Equipment 167
Calibration of Shrinkage Limit Dish 167
Procedure 168
Calculations 170
Example 170
Discussion 171
Shrinkage Limit by Wax Method 171

Introduction 171
Standard 171
Equipment 172
Calibration of Shrinkage Limit Dish 172
Procedure 173
Calculations 174
Discussion 175
Shrinkage Limit by Direct Measurement Method 175
Introduction 175
Standard 175
Equipment 175
Procedure 176
Calculations 176
Example 177
Discussion 178
References 178
Additional Reading 179

14 Linear Shrinkage   **181**

Introduction 181
Standard 181
Equipment 182
Procedure 182
Calculations 184
Example 185
Discussion 185
Linear Shrinkage by Coefficient of Linear Extensibility
  (COLE): 187
References 188
Additional Reading 189

15 Shrink Test   **191**

Introduction 191
Standard 191
Equipment 192
Procedure 192
Calculations 194
Example 194
Discussion 196
References 196
Additional Reading 196

16 Free-Swell Index Tests 197

Background 197
Free-Swell Index (FSI) 198
  Introduction 198
  Standard 199
  Equipment 199
  Procedure 200
  Calculations 200
  Example 202
  Discussion 202
Differential Free-Swell Index (DFSI) and Free-Swell Ratio
  (FSR) 203
  Introduction 203
  Standard 203
  Equipment 203
  Procedure 204
  Calculations 204
  Example 204
  Discussion 205
References 207
Additional Reading 207

17 Dispersion 209

Background 209
Dispersion by Double Hydrometer Test 210
  Introduction 210
  Standard 210
  Equipment 210
  Procedure 210
  Calculations 211
  Example 212
  Discussion 212
Dispersion by Pinhole Test 212
  Introduction 212
  Standard 213
  Equipment 213
  Procedure 214
  Calculations 215
  Example 215
  Discussion 215
Dispersion by Crumb Test 218
  Introduction 218

Standard 218
Equipment 218
Procedure 218
Calculations 219
Discussion 219
References 219
Additional Reading 220

18 Initial Consumption of Lime   221

Background 221
Introduction 222
Standard 222
Equipment 222
Procedure 222
Calculations 223
Example 223
Discussion 224
Reference 225
Additional Reading 225

19 Lime Fixation Point   227

Introduction 227
Standard 227
Equipment 227
Procedure 228
Calculations 228
Example 228
Discussion 229
References 230
Additional Reading 230

20 Hygroscopic Water Content   231

Introduction 231
Standard 231
Equipment 231
Preparation of Humidity Chamber 232
Procedure 232
Calculations 232
Example 233
Discussion 233
References 235
Additional Reading 236

21 Soil Suction Using Filter Paper   237

Introduction 237
Standard 237
Equipment 238
Filter Paper Calibration 238
Procedure 239
Calculations 240
Example 241
Discussion 241
References 243
Additional Reading 243

22 Thermal Conductivity   245

Introduction 245
Standard 246
Equipment 247
Calibration of Thermal Needle 247
Procedure 247
Calculations 248
Example 248
Discussion 249
References 250
Additional Reading 252

23 Electrical Resistivity   255

Introduction 255
Standard 258
Equipment 258
Calibration 258
Procedure – Four Electrode Method 259
Calculations 260
Procedure – Two Electrode Method: 260
Calculations 260
Discussion 261
References 266
Additional Reading 268

24 Index Strength Tests and Sensitivity   271

Introduction 271
Pocket Penetrometer 271
  Standard 272

Equipment 273
Procedure 273
Calculations 275
Pocket Hand Vane (Torvane) 276
Introduction 276
Standard 277
Equipment 277
Procedure 278
Calculations 279
Laboratory Vane 279
Introduction 279
Standard 281
Equipment 281
Procedure 282
Calculations 283
Fall Cone 284
Introduction 284
Standard 284
Equipment 284
Procedure 285
Calculations 287
Sensitivity 288
Sensitivity and Liquidity Index 288
Discussion 290
References 292
Additional Reading 294

25 **Thixotropy**   295

Introduction 295
Standard 296
Equipment 297
Procedure 297
Preparation of Test Specimens 297
Measurement of Undrained Shear Strength 299
Calculations 300
Discussion 300
References 301
Additional Reading 302

Index   305

Preface

Most geotechnical site characterization studies of fine-grained soils involve laboratory testing that includes traditional evaluation of basic soil behavior, such as Atterberg Limits, undrained shear strength and consolidation. In most cases, there is little attention given to the determination of other soil Compositional Characteristics that might be used to characterize the soil in more detail, to distinguish subtle changes in soil conditions that may show vertical variations in soil that might not otherwise be obvious. Several recent papers have presented recommendations for characterizing fine-grained soils (e.g., Ladd & DeGroot 2008); however, these recommendations do not include consideration of the Compositional Characteristics of the soil. In many cases, the Compositional Characteristics of fine-grained soils are even omitted in descriptions of soil behavior. At most, the basic soil characterization only includes Atterberg Limits in order to use a Casagrande Plasticity Chart to classify the soil according to the Unified Soil Classification System.

There may be good reasons to go beyond this basic approach and use more advanced laboratory tests to provide determination of more detailed Compositional Characteristics. Most fine-grained soil deposits, whether derived from alluvial, marine, lacustrine, eolian or glacial processes or residual soils, have characteristics largely resulting from the parent material, depositional environment or degree of weathering. Because most geologic deposits accrue over time, there may be variations in composition at a single location, produced simply by variations in parent material and other factors over time. An individual deposit may appear to be more or less homogeneous but may show variations in composition that can be determined using other characterization tests.

Advanced laboratory characterization may provide information that helps explain unusual behavior observed in other tests that are often not easily explained. There are other simple and advanced characterization tests that might be considered for inclusion in the soil characterization for some site investigations. They include determining Carbonate Content, Specific Surface Area, Cation Exchange Capacity, Pore Fluid Salinity and other tests.

Geotechnical laboratory tests can generally be placed into one of three categories: 1) characterization tests, 2) behavioral tests, or 3) structural property tests.

Characterization Tests are those tests that are performed to determine soil composition and quantify the individual components of a soil. Essentially, they are conducted to determine soil particle characteristics. Examples include grain-size distribution and Carbonate Content.

Behavioral Tests are those tests that are performed to determine how the soil behaves under certain conditions. Examples are the Atterberg Limits tests, which demonstrate how the fines fraction of a soil reacts when mixed with water; or the Linear Shrinkage test, which demonstrates how much a soil shrinks when water is removed from it. In these tests, we are observing and quantifying the behavior of the soil.

Structural property tests are those tests that are performed to obtain values of specific properties of soils that will most likely be used in design. Examples of these types of tests are the triaxial compression test, which provides an indication of the strength of the soil; or the one-dimensional consolidation test, which provides a measure of the soil stress history and compressibility.

This manual covers soil characterization and behavioral tests of fine-grained soils. It is intended to be used in a graduate level university course in Soil Mechanics/Soil Behavior but should also be of interest to researchers and practicing engineers.

Acknowledgments

The author would like to acknowledge the contribution of many colleagues and graduate students over the author's 36-year academic career at Iowa State University, Clarkson University and the University of Massachusetts, who have contributed to various aspects of laboratory testing. The author would also like to thank Professor Don DeGroot, who provided valuable comments to the author on an early draft of this manual.

Introduction

Almost every geotechnical site investigation includes sampling of soil and rock and some level of laboratory testing to evaluate specific characteristics.

SAMPLES VS. SPECIMENS

Samples are generally collected during a field investigation to be representative of individual geologic strata at specific locations and depths. Samples may be *undisturbed*, as taken from thin-walled Shelby tubes or other special samplers; or they may be *disturbed*, as collected from a split spoon drive sampler used with the standard penetration test or collected from auger cuttings as a borehole proceeds. Disturbed samples may also be taken in bulk quantity, as from an excavation or road cut.

Specimens are taken from samples for specific tests and are essentially a subset of the sample. They are often carefully trimmed in the laboratory for use in strength or stiffness tests or they may be taken from a disturbed sample to determine Water Content, grain-size distribution or Atterberg Limits. Samples are intended to be representative of the strata. Specimens are intended to be representative of the sample.

DIRECT VS. INDIRECT TESTS

There are different types of laboratory tests for soils. Most tests are either *direct* or *indirect*. *Direct* tests involve the direct determination of the characteristic to be obtained from the test. For example, determination of Water Content by convection oven (Chapter 1) involves the direct measurement of soil mass before and after oven drying. These mass values and calculated mass of water removed by drying are used to calculate soil Water Content. The only measurements required are mass values, typically obtained from a calibrated electronic laboratory scale. The same may be said for Water Content determined by the microwave oven method.

On the other hand, Water Content determined by the gas carbide method (Chapter 1) first involves the measurement of gas pressure in a closed pressure vessel and relies on the chemical reaction between calcium carbide and soil water. The gas pressure is then converted to soil Water Content using a calibration chart prepared using soil with known values of Water Content for that particular pressure vessel. This is an *indirect* test method.

Similarly, the determination of Carbonate Content using the Rapid Carbonate Analyzer (Chapter 6) involves the measurement of pressure in a closed pressure chamber and relies on the chemical reaction between dilute hydrochloric acid and carbonates present in the soil. The measured gas pressure is then converted to soil Carbonate Content using a calibration chart prepared using known quantities of calcium carbonate for that particular pressure chamber.

Since many tests are indirect tests, this means that careful calibrations are essential to the quality of the determination of soil characteristics. Calibrations should be performed with both accuracy and the required precision needed for a particular procedure. Occasionally, calibrations should be checked to make sure that the equipment is still performing correctly.

TESTING STANDARDS

Standard procedures for testing soils have been developed over the years so that the same procedures and equipment are used throughout the profession. This allows test results obtained from one individual to be compared to similar tests obtained by another individual, around the globe, without issues related to how a particular test may have been performed. There are a number of standardization agencies that have been in operation since the 1940s, including the ASTM (American Society for Testing and Materials), AASHTO (American Association of State Highway and Transportation Officials) and ISO (International Standards Organization). In addition, many state highway departments in the U.S. have developed internal testing standards to suit local geologic conditions and problems within their state. Many countries have also developed testing standards that may include tests not covered by ASTM or AASHTO, for example British Standards (BS) in the UK and Indian Standards in India.

In this manual, reference is made in each chapter to the current ASTM standards for individual tests, where appropriate. However, several of the chapters cover tests that are not currently standardized. In these cases, the test procedure described is the most current method used, and cited references have been provided referring to the test procedure recommended by well-documented research. An additional reading list of references related to the test described in the chapter is also given.

CHARACTERISTIC VS. PROPERTY

The term *Property* is often used to describe specific attributes of soils. In fact, the term is often misused, since there are very few true soil properties. Sometimes, a better term to use might be *characteristic* or *behavior*. The use of *Property* should be restricted to those attributes of the soil that are intrinsic and do not change over time or do not depend on the test method used to obtain the outcome. As an example, the undrained shear strength of saturated clays is not truly a soil property since different values may be obtained from different test methods, e.g., field vane, simple shear, unconfined compression, Fall Cone, etc., even though they all may be standardized. Even using the same test method may give a different outcome, for example if the strain rate or rate of loading is changed. On the other hand, the grain-size distribution is considered a Soil Property (in this manual, it is referred to as a *Compositional Characteristic*). It does not change with time, unless the soil is washed and some fines are lost or unless soil is added. It is considered intrinsic to the soil.

Determination of Compositional Characteristics

Part 1

Determination
of Compositional
Characteristics

Chapter 1

Water Content

BACKGROUND

Soils are composed of three phases: solid minerals, water and air. The amount of water in soil can significantly influence soil behavior, especially for fine-grained soils such as clays, silty clays and silts. Water Content is a fundamental soil characteristic and should be performed on every sample collected as part of a project. The test often uses cuttings from specimens trimmed in the lab for other tests, such as consolidation or direct shear. Water Content of soils is a mass relationship and is defined as the ratio of the mass of water to the mass of oven dry solids expressed as a percent. The test is also required as part of the compaction test (not covered in this manual) and is also conducted on samples collected in the field as part of a drilling and sampling program.

There are three common laboratory methods used to determine Water Content of soils: 1) convection oven, 2) microwave oven and 3) gas carbide. The first two methods are used exclusively in the laboratory, while the gas carbide method may be used in the laboratory or in the field. Another method not covered in this manual is the determination of Water Content by direct heating by using a hot plate, heat lamp, gas stove, etc. This is covered by ASTM D4959 *Standard Test Method for Determination of Water (Moisture) Content of Soil by Direct Heating Method.*

WATER CONTENT BY CONVECTION OVEN

Introduction

In most geotechnical laboratories, the routine determination of soil Water Content is performed using a convection oven. Wet samples are placed in a clean dry bowl or aluminum moisture tin (called a "tare") of known mass. The total mass of the wet soil and bowl is determined, and then the bowl and soil are placed in the oven to dry at $110° \pm 5°C$ until a constant mass is achieved. Many labs leave the soil in the oven to dry for 24 h as a matter of

DOI: 10.1201/9781003263289-2

3

convenience. After drying, the soil and bowl are removed and the dry mass is determined. The test is simple and inexpensive and only requires basic laboratory equipment. Figure 1.1 shows examples of ceramic and aluminum moisture tares on an electronic scale. Figure 1.2 shows samples drying in a convection oven.

Figure 1.1 Examples of ceramic and aluminum moisture tins.

Figure 1.2 Soil samples drying in a convection oven.

Standard

ASTM D2216 *Standard Test Method for Laboratory Determination of Water (Moisture) Content of Soil and Rock by Mass*

Equipment

Convection oven capable of maintaining 110°±5°C
Aluminum tins (tare) or ceramic bowls
Electronic scale with a minimum capacity of 500 gm and a resolution of 0.01 gm
Spatula
Tongs or heat proof gloves
Desiccator

Procedure

1. Determine the mass of a clean dry ceramic bowl or aluminum moisture tin (M_{TARE}) to the nearest 0.01 gm. (It is best to select a bowl or tin with a smaller mass than the wet soil to reduce errors.)
2. Select a representative sample of soil and place it in the bowl or tin and determine the wet mass of soil and bowl ($M_{WET} + M_{TARE}$).
3. Place the bowl or tin uncovered in a convection oven set at 110°±5°C until the dry mass is constant.
4. After drying, carefully remove the bowl with soil from the oven and place the bowl in a desiccator to cool. As soon as the bowl has cooled, immediately determine the mass of dry soil and bowl ($M_{DRY} + M_{TARE}$).

Calculations

Water Content is expressed as the mass of water divided by the oven dry mass of soil in percent:

$$W = (M_{WATER}/M_{DRY}) \times 100\% \tag{1.1}$$

where:
W = Water Content
M_{WATER} = mass of water = $(M_{WET} + M_{TARE}) - (M_{DRY} + M_{TARE})$
M_{DRY} = mass of oven dry soil = $(M_{DRY} + M_{TARE}) - M_{TARE}$

Example

A wet sample of soil was placed in a dry ceramic bowl with a mass of 126.51 gm. The total mass of the wet soil and bowl was 288.62 gm. After oven drying for 24 h, the mass of the dry soil and bowl was 262.18 gm. Determine the Water Content.

$$\text{Mass of Water} = M_{WATER} = (M_{WET} + M_{BOWL}) - (M_{DRY} + M_{BOWL})$$

$$= 288.62\,\text{gm.} - 262.18\,\text{gm} = 26.44\,\text{gm}$$

$$\text{Mass of Dry Soil} = (M_{DRY} + M_{BOWL}) - M_{BOWL}$$

$$= 262.18\,\text{gm} - 126.51\,\text{gm} = 135.67\,\text{gm}$$

$$W = (M_{WATER}/M_{DRY}) \times 100\% = (26.44\,\text{gm} / 135.67\,\text{gm}) \times 100\% = 19.5\%$$

It is routine practice to report Water Content to the nearest 0.1%. Never report Water Content to the nearest 0.01%; this gives a false sense of precision.

Discussion

Determining Water Content of soils is easy, but errors can occur. Care should be taken when placing the bowl or tin into or removing it from the oven so that no soil is spilled. The bowl should be large enough so that the soil can be spread evenly around the base and not heaped up in the bowl. This will help the drying process. Bowls should be clean and dry before use and should be numbered or lettered to keep track of which sample is which. Be sure that the scale has been recently calibrated and properly zeroed before use. Never move a scale from the lab bench; it should remain in a fixed position all the time. It is best to use aluminum or stainless-steel moisture tins and ceramic bowls since they won't rust over time.

How much soil should be used to obtain a representative determination of Water Content? ASTM D2216 gives recommended minimum amounts of wet soil mass that should be used, depending on the grain size of the soil. These are summarized in Table 1.1. For fine-grained soils, a lower limit is set at 20 gm.

The dry mass of the oven-dried soil and bowl needs to be determined quickly when removed from the desiccator. Moisture from the atmosphere can be adsorbed by the soil, especially clays, if left out in the open. The amount of soil adsorbed depends on the relative humidity in the laboratory, time and the soil. Adsorbed water will change the dry mass and introduce an error in the measurements and calculations. How much moisture can a sample adsorbed if the sample is left out too long? Figure 1.3 shows results of mass measurements with time for two soil samples that were removed from the oven but then left out on the lab bench for several hours. Although the increase in mass is small, it is nonetheless an error that should be avoided.

Table 1.1 Recommended Minimum Wet Soil Mass for Determining Water Content
(after ASTM D2216)

Maximum particle size (mm)	Standard sieve size	Recommended minimum mass of wet soil for Water Content reported to nearest 1%	Recommended minimum mass of wet soil for Water Content reported to nearest 0.1%
≤2	No. 10	20 gm	20 gm
4.75	No. 4	20 gm	100 gm
9.5	3/8 in	50 gm	500 gm
19.0	¾ in	250 gm	2.5 kg
37.5	1 ½ in	1 kg	10 kg
75	3 in	5 kg	50 kg

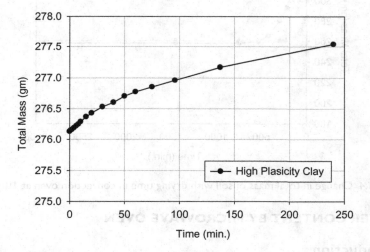

Figure 1.3 Increase in mass of dry soil when left out in the open after oven drying.

Does the soil really need to dry in the oven overnight or for 24 h? ASTM D2216 does not actually require this drying time; it only states that

The time required to obtain constant mass will vary depending on the type of material, size of specimen, oven type and capacity, and other factors. The influence of these factors generally can be established by good judgement, and experience with the materials being tested and the apparatus being used.

and

In most cases, drying a test specimen overnight (about 12 to 16 hrs.) is sufficient. In cases where there is doubt concerning the adequacy of drying, drying should continue until the change in mass after two successive periods (greater than 1 hr.) of drying is an insignificant amount (less than about 0.1%). Specimens of sand may often be dried to constant mass in a period of about 4 hrs., when a forced-draft oven is used.

Figure 1.4 shows the change in mass of several soils with time in the oven. Even though it appears that less time could be used, it is very common routine laboratory practice to allow soil samples to dry for 24 h to determine Water Content.

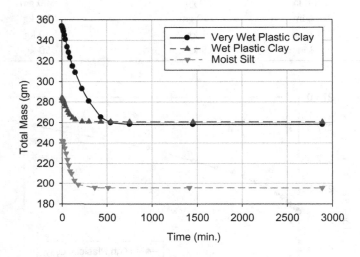

Figure I.4 Change in total mass of soil with drying time in convection oven at 110°C.

WATER CONTENT BY MICROWAVE OVEN

Introduction

For rapid laboratory determination of soil Water Content, a technique using a microwave has been around since the 1970s and can eliminate the waiting periods typically used with the convection oven method. The procedure is similar to the convection oven procedure described in Section 1.1; that is, the water is driven from the sample leaving only dry soil. While the convection oven method described in Section 1.1 is preferred for most routine laboratory work, the Microwave oven method can be useful and was introduced to reduce the time to obtain Water Content. The two tests give very similar results in most soils; however, this test method should not be used with highly organic soils.

Standard

ASTM D4643 *Standard Test Method for Determination of Water (Moisture) Content of Soil by Microwave Oven Method*

Equipment

Microwave oven (power rating in the range of 700 watts)
Ceramic bowls
Electronic scale with a minimum capacity of 250 gm and a resolution of
  0.01 gm
Tongs or heat-proof gloves

Procedure

1. Determine the mass of a clean dry ceramic bowl, M_{BOWL}, to the nearest 0.01 gm. (Note: Do not use metal tares.)
2. Place a representative sample of the moist soil in the bowl, spread the soil evenly in the bowl and determine the mass of the soil and bowl, $(M_{WET} + M_{BOWL})$, to the nearest 0.01 gm.
3. Place the bowl and soil in a microwave oven and dry on "high" setting for 3 min.
4. After 3 min, remove the bowl and determine and record the mass, $(M_{DRY} + M_{BOWL})$, to the nearest 0.01 gm.
5. Return the bowl and soil to the oven and repeat the drying for 1 min. Remove the bowl and soil from the oven and determine and record the mass.
6. Repeat the procedure of drying for a duration of 1 min until the change in mass between two consecutive measurements is insignificant, generally less than 0.1 % Water Content.

Calculations

Determine the Water Content as:

$$W = (M_{WATER}/M_{DRY}) \times 100\% \tag{1.2}$$

where:
  W = Water Content
  M_{WATER} = mass of water = $(M_{WET} + M_{BOWL}) - (M_{DRY} + M_{BOWL})$
  M_{DRY} = mass of oven dry soil = $(M_{DRY} + M_{BOWL}) - M_{BOWL}$

Example

A moist sample of soil was placed in a dry ceramic bowl with a mass of 108.22 gm. The initial total mass of the wet soil and bowl was 238.57 gm. After 3 min in the microwave oven, the mass of the soil and bowl was 217.36 gm. The bowl was placed back in the microwave. After 1 min more in the microwave oven, the mass of the soil and bowl was 217.22 gm. The bowl was placed back in the microwave. After one more minute in the microwave oven, the mass of the soil and bowl was again 217.22 gm. Determine the Water Content.

$$\text{Mass of Water} = M_{WATER} = \left(M_{WET} + M_{BOWL}\right) - \left(M_{DRY} + M_{BOWL}\right)$$

$$= 238.57\,\text{gm.} - 217.22\,\text{gm} = 21.35\,\text{gm}$$

$$\text{Mass of Dry Soil} = \left(M_{DRY} + M_{BOWL}\right) - M_{BOWL}$$

$$= 217.22\,\text{gm} - 108.22\,\text{gm} = 109.00\,\text{gm}$$

$$W = \left(M_{WATER}/M_{DRY}\right) \times 100\% = \left(21.35\,\text{gm}/109.00\,\text{gm}\right) \times 100\% = 19.6\%$$

Discussion

How much mass should be used for the Microwave method? ASTM D4643 gives recommended mass of the specimen based on different particle sizes, Table 1.2.

How much time is needed to completely dry a sample using the microwave oven method? Unlike the convection oven method, which may take up to 12 h to completely dry a specimen, the microwave oven method will typically dry small mass specimens within about 6–8 min; longer times will be needed for large masses of soil. Figure 1.5 shows the total mass of soil + water for several samples of fine-grained soil after different drying times. Each sample started with the same dry mass but different initial Water Content. When the total mass reaches a constant value, this indicates that all of the water has been removed and the specimen is dry. In low Water Content soils, even clays, this only takes 2–3 min. In very high Water Content soils, drying is essentially complete after about 7 min.

Does the microwave oven method give the same results as the convection oven method? Several studies have shown that the Water Content determined by both the convection oven and microwave oven methods are essentially the same (Lade & Nejadi-Babadai 1976; Charlie et al. 1982; Carter & Bentley 1986; Hagerty et al. 1990; Gilbert 1991; Mendoza & Orozco 1999). The microwave oven method is an expedient means of determining

Table 1.2 Recommended Soil Mass for Water Content Determination

Sieve retaining not more than about 10% of soil	Soil types	Wet soil mass
No. 10 (2.0 mm)	Med. to fine sands, silts, clays	100–200 gm
No. 4 (4.75 mm)	Fine gravel and coarse sand	300–500 gm
¾ in. (19 mm)	Gravel	500–1000 gm

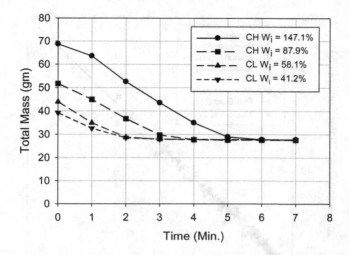

Figure 1.5 Time needed to obtain constant mass in microwave oven for a high plasticity clay (CH) and low plasticity silty clay (CL) at two different initial water contents.

Water Content and is especially useful for coarse-grained and mixed-grain soils. For fine-grained soils, especially clays, the process may require several sequences of drying and weighting to obtain a final dry mass but still can usually determine Water Content within about 10 min. The method is useful when performing laboratory compaction tests to determine if soil is mixed to the target Water Content. Figure 1.6 shows a comparison between Water Content determinations made by the author using the convection oven method and the microwave oven method for a number of different soils.

Do the results from the microwave oven method and convection oven method show similar variability? Table 1.3 shows results of replicate Water Content samples taken from two bulk samples of a clayey silt prepared at two different Water Contents for a laboratory compaction test. Water Content was determined on 20 samples taken from the bags using both the convection oven and the microwave oven methods using the procedures previously described. These results show that there is no statistical difference in the mean values of Water Content obtained between these two methods. The coefficient of variation is all less than 10%.

WATER CONTENT BY PORTABLE CARBIDE GAS METHOD – "SPEEDY" MOISTURE METER

Introduction

It is sometimes necessary or convenient to obtain rapid measurements of soil Water Content in the field. For example, during compaction of soils for

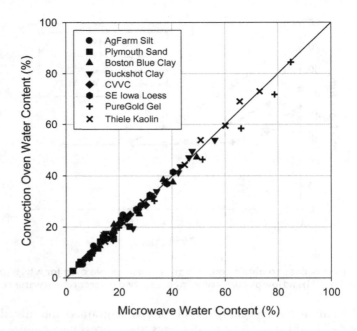

Figure 1.6 Comparison between convection oven method and microwave oven method for a number of fine-grained soils.

Table 1.3 Variation between Convection Oven and Microwave Oven Methods

	Convection oven			Microwave oven		
	Mean Water Content (%)	Standard deviation (%)	Coefficient of variation (%)	Mean Water Content (%)	Standard deviation (%)	Coefficient of variation (%)
Sample 1	15.7	0.7	4.2	15.5	1.2	7.7
Sample 2	21.7	0.6	2.8	21.6	0.6	2.8

construction of an embankment or an earth dam or placement of backfill under a slab or behind a retaining wall. The contractor may need a determination of Water Content to see if the Water Content of the soil is within the compaction specifications. A simple and fast method that has proven to be reliable in a wide range of soils is based on the calcium carbide gas tester, available commercially as the Speedy Moisture Meter, which refers to the device. The gas carbide method is generally applicable to most soils but is restricted to soils with particles no larger than the No. 4 sieve (0.425 mm).

The Speedy was developed in England in the early 1960s and is still in use extensively around the world. The device may be used in the laboratory or in the field, is completely portable and requires no power. The method is based on the concept that when dry calcium carbide is mixed with moist soil, the water in the soil reacts with the calcium carbide to produce acetylene gas. This reaction is given by:

$$CaC_2 + 2H_2O \text{ gives } Ca(OH)_2 + C_2H_2 \tag{1.3}$$

The mixing is performed in a closed pressure vessel, and the generated gas pressure inside the vessel is read directly on a pressure gauge; the more the moisture, the more the gas pressure. In order to produce reliable results, the carbide reagent must mix with all of the water in the soil. The Water Content is obtained using a calibration chart or equation, either provided by the manufacturer or developed by the operator. Figure 1.7 shows a photo of the traditional Speedy equipment provided in a wooden case with pressure vessel, scale, scoop, brush and steel balls.

Even though a calibration chart to convert the Speedy gauge readings to oven dry Water Content is usually provided by the manufacturer, it is suggested that a calibration chart be developed for each device. This is done by preparing several soil specimens at different Water Content and determining the oven dry Water Content and Speedy gauge reading on replicate specimens.

Standard

ASTM D4944 *Standard Test Method for Field Determination of Water (Moisture) Content of Soil by the Calcium Carbide Gas Pressure Tester*

Figure 1.7 Original Speedy moisture tester kit in wood case.

Equipment

Speedy Moisture Meter

Scale (either provided with Speedy equipment or small portable electronic laboratory scale with a minimum capacity of 250 gm and a resolution of 0.1 gm)

Small spatula

Two 32 mm (1.25 in) steel balls

Cleaning brush

Calibration chart

Calcium carbide reagent

Procedure

1. Inspect the Speedy Moisture Meter to make sure that it is clean and dry. Record the serial number. Hold the Speedy in the horizontal position and read the pressure gauge. If the gauge does not read zero, record the reading as R_0.
2. Remove the cap from the Speedy and inspect the rubber gasket around the cap. Apply a thin film of silicone grease around the gasket so that a good seal will be created during the test.
3. Use the Speedy scale or an electronic scale to weigh out 26 gm of wet soil for the test specimen.
4. Place the soil specimen into the cap of the Speedy.
5. Place three scoops of calcium carbide reagent into the body of the Speedy. Carefully place the two steel balls into the Speedy.
6. Hold the Speedy in the horizontal position, place the cap onto the Speedy and tighten the clamp. (Note: When placing the cap on, be sure that soil does not spill into the Speedy and come into contact with the reagent until the clamp has been tightened.)
7. Now raise the Speedy to the vertical position with the pressure gauge facing down so that soil from the cap mixes with the calcium carbide reagent.
8. Hold the Speedy in the horizontal position with the pressure gauge facing the operator and gently roll the Speedy back and forth so that the steel balls and reagent mix with the soil and expose all of the soil moisture to the reagent. (Note: Do not shake the Speedy up and down, as this could damage the pressure gauge.)
9. Periodically check the pressure gauge. Usually, the gauge will increase quickly as the internal reaction between the soil water and calcium carbide begins but will eventually slow down and reach a maximum value. This typically occurs in about 1–2 min for sandy soils and about 2–4 min for clays.
10. When the gauge reading on the Speedy becomes constant, record the final gauge pressure reading as R_F.

11. Hold the Speedy away from the operator and loosen the clamp. This will release the gas pressure. Empty the contents into an appropriate receptacle and retrieve the steel balls. Use the brush to clean out the Speedy and the cap so that they are ready for the next test.

Calculations

Determine the corrected pressure reading as:

$$R_C = R_F - R_0 \tag{1.4}$$

There are no other calculations with this test. The Water Content (%) is obtained by taking the corrected reading R_C and converting the reading to % Water Content using the calibration chart or using a calibration regression equation.

For very wet soils, the pressure may go beyond the limit of the Speedy pressure gauge. In this case, an initial soil mass of 13 gm should be used instead of 26 gm. When half of the wet soil mass is used, the final pressure gauge reading must be multiplied by two to give the correct pressure before determining Water Content.

Example

Speedy No. 30477 was used to determine the Water Content of a stiff clay sample. An initial gauge reading of $R_0 = 0.6$ and a final gauge reading of $R_F = 12.8$ were obtained using a 26 gm sample. The calibration chart for this instrument is shown in Figure 1.8. Determine the Water Content.

The corrected gauge reading is: $R_c = R_f - R_0 = 12.8 - 0.6 = 12.2$
Using the calibration chart, the Water Content is: $W = 12.2 \times 1.07 = 13.1\%$

Discussion

Operators may wish to use a dust mask and disposable laboratory gloves to perform the test to reduce contact with the gas and calcium carbide reagent. The Speedy is ideally suited to determining Water Content in the field since it requires no power and is portable. A number of laboratory studies have shown that the Speedy provides reliable measurements of Water Content for a wide range of soils when used properly and when properly calibrated (Blystone et al. 1961; Schwartz 1966; Smith 1970; Berney et al. 2012; Arsoy et al. 2014). The calibration chart shown in Figure 1.8 was obtained by the author using four different soils, ranging from low plasticity silt to high plasticity clay, and indicates that the Speedy does not provide a direct reading of the Water Content. Note that all of the calibration points obtained using duplicate samples fall over the 1:1 line and show that for this device (No. 30477) the calibration is: $W = 1.07 \times R_C$.

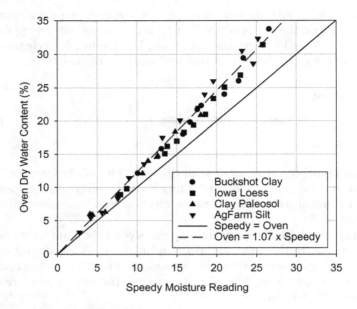

Figure 1.8 Calibration of Speedy Moisture Meter with different soils.

Table 1.4 Variability of Convection Oven, Microwave Oven and Speedy Moisture Meter for Three Soils at Two Different Water Content Levels

	Mean W (%)	Standard deviation (%)	Coefficient of variation (%)
Convection oven	15.1	0.5	3.5
Microwave oven	14.6	0.4	2.7
Speedy	14.8	0.3	2.0
Convection oven	45.4	0.2	0.5
Microwave oven	45.6	0.3	0.8
Speedy	45.5	1.2	2.6

Each Speedy should be calibrated individually so that any differences in manufacturing can be accounted for using a specific calibration chart. How repeatable are the values obtained from the Speedy? Are results less reliable than either convection oven method or microwave oven method values of Water Content? Table 1.4 shows results obtained using large patches of soil prepared to different Water Contents. Twenty tests were performed with the convection oven, microwave oven, and Speedy methods at two different Water Contents (a total of 120 tests). The results show that in every case, the coefficient of variation is less than 5%, which is very good for soil tests, considering the possibility of natural variations in the soil.

Figure 1.9 New generation Speedy Moisture Tester and plastic case with electronic scale.

Speedy moisture testers are now available from a number of manufacturers and are very compact. A new generation Speedy, with a small portable battery-powered electronic scale and hard plastic case, is shown in Figure 1.9. Some models are even equipped with a digital pressure gauge connected to the pressure vessel to reduce operator error in reading the analog pressure gauge.

REFERENCES

Arsoy, S., Keskin, E., and Ozgur, M., 2014. Reliability of Soil Water Content Measurements by the Calcium Carbide Gas Pressure Method for Small Specimens. *Scientia Iranica*, Vol. 21, No. 6, pp. 1762–1772.

Berney, E.S. IV, Kyzar, J.D., and Oyelami, L.O., 2012. Device Comparison for Determining Field Soil Moisture Content. US Army Corps of Engineers Report ERDC/GSL TR-11-42.

Blystone, J.R., Pelzner, A., and Steffens, G.P., 1961. Moisture Content Determination by the Calcium Carbide Gas Pressure Method. *Public Roads*, Vol. 31, No. 8, pp. 177–180.

Carter, M., and Bentley, S.P., 1986. Practical Guidelines for Microwave Drying of Soils. *Canadian Geotechnical Journal*, Vol. 23, pp. 598–601.

Charlie, W.A., Von Gunten, M.W., and Doehring, D.O., 1982. Temperature Controlled Microwave Drying of Soils. *Geotechnical Testing Journal, ASTM*, Vol. 5, No. 3/4, pp. 66–75.

Gilbert, P.A., 1991. Rapid Water Content by Computer Controlled Microwave Drying. *Journal of Geotechnical Engineering, ASCE*, Vol. 117, No. 1, pp. 118–138.

Hagerty, D.J., Ullrich, C.R., and Callan, C.A., 1990. Microwave Drying of Highly Plastic and Organics Soils. *Geotechnical Testing Journal, ASTM*, Vol. 13, No. 2, pp. 142–145.

Lade, P.V., and Nejadi-Babadai, H., 1976. Soil Drying by Microwave Oven. *ASTM STP*, Vol. 599, pp. 320–340.

Mendoza, M.J., and Orozco, M., 1999. Fast and Accurate Techniques for Determination of Water Content in Soils. *Geotechnical Testing Journal, ASTM*, Vol. 322, No. 4, pp. 301–307.

Schwartz, A.E., 1966. Rapid Means of Determining Density and Moisture Content of Soils and Granular Materials. Final Report PB173170, Department of Civil Engineering, Clemson University.

Smith, P.C., 1970. Suggested Methods for Rapid Determination of Approximate Moisture Content of Soils for Field Control of Embankment Construction. *ASTM STP*, Vol. 479, pp. 447–459.

ADDITIONAL READING

Gee, G.W., and Dodson, M.E., 1981. Soil Water Content by Microwave Drying: A Routine Procedure. *Soil Science Society of America Journal*, Vol. 45, No. 6, pp. 1234–1237.

Hagerty, D.J., Ullrich, C.R., and Denton, M.M. 1990. Microwave Drying of Soils. *Geotechnical Testing Journal, ASTM*, Vol. 13, No. 2, pp. 138–142.

Kramarenko, V.V., Nikitenkov, A.N., Matveenko, I.A., Molokov, Y., and Vasienko, Y.S., 2016. Determination of Water Content in Clay and Organic Soil Using Microwave Oven. *Earth and Environmental Science*, Vol. 43, pp. 1–6.

Mendoza, M.J., 1992. Discussion of Rapid Water Content by Computer-Controlled Microwave Drying. *Journal of Geotechnical Engineering, ASCE*, Vol. 117, No. 1, pp. 118–138, pp. 1129–1131.

Miller, R.J., Smith, R.B., and Biggar, J.W., 1974. Soil Water Content: Microwave Oven Method. *Soil Science Society of America Proceedings*, Vol. 38, No. 3, pp. 535–537.

Ryley, M.D., 1969. The Use of a Microwave Oven for the Rapid Determination of Moisture Content of Soils. Road Research Laboratory Report No. LR280, Crowthorne, England.

Chapter 2

Phase Relationships of Soils

Determining Unit Weight, and Calculating Voids Ratio, Porosity and Saturation

BACKGROUND

An important component of nearly every geotechnical investigation involves determining basic engineering characteristics of soil samples. Undisturbed core samples are often obtained from test borings in the field, and these samples are used to determine Unit Weight and Water Content. From these measurements and from an estimate or measurement of the density of solids, ρ_S (Specific Gravity), the Voids Ratio, Porosity and Saturation may be calculated. These quantities are used routinely by geotechnical engineers and are a fundamental way of describing the state of the soil.

Soils can be thought of as a composite material consisting of three different components or phases: 1) solid soil mineral matter; 2) water; and 3) air. Sometimes a soil consists of only the solid and water components and there is no air, i.e., all of the space between solid particles is occupied by water, for example, below the water table. It is important to quantify the different amounts of each of these components in any given soil and therefore we need specific definitions for both mass amounts and volume amounts of each component. Most core samples obtained from test borings are cylindrical, but sometimes cubical specimens are hand carved from excavations, test pits or road cuts; in either case, however, the procedure is the same. We need to determine the total volume and total mass of the specimen and then determine the Water Content using one of the methods described in Chapter 1.

Figure 2.1 shows a cross section of an idealized soil block of unit volume and illustrates the way in which we can visualize the different components in the soil. Note that this discussion is related only to mass and volume and does not consider the different constituents within the solid phase (such as mineralogy) that might influence soil behavior.

DOI: 10.1201/9781003263289-3

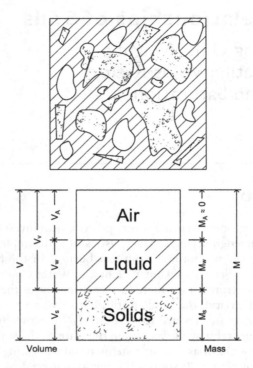

Figure 2.1 Phase composition of soil.

PHASE RELATIONSHIPS IN SOILS

Phase relationships based on mass

Water Content

The only relationship between the various phases of soil based solely on *mass* composition is the Water Content, described in Chapter 1 and defined as the mass of water divided by the mass of solids, expressed as a percentage:

$$\text{Water Content} = W = (M_W/M_S) \times 100\% \tag{2.1}$$

where:
  M_w = Mass of water
  M_s = Mass of solids

There are several methods for determining Water Content as described in Chapter 1. The most common method is the convection oven method. Based on Equation 2.1, the range of Water Content of soils could be from 0% for a completely oven dry soil to well over 100% for a very wet soil. In the field, there is really no such thing as a truly "dry" soil, although the term "dry" is often used when creating a soil description when soils have very low Water Content, such as in arid environments. Because the definition of Water Content is based on the ratio of two quantities; if the mass of solids is very low, for example as with highly organic soils below the water table, the Water Content could easily be 400% or higher. For most soils, however, the natural Water Content typically ranges from about 5%–10% for soils in dry desert or arid environments to about 100%–200% for off-shore marine deposits.

As noted in Chapter 1, Water Content is one of the most useful quantities for soils and should be routinely measured on all samples taken on projects. Water Content of fine-grained soils is used to place them in the context of soil plasticity, i.e., Liquidity Index, as will be discussed in Chapter 12.

Phase relationships based on volume

As previously noted and shown in Figure 2.1, a unit volume of soil is made up of either the solid phase and the non-solid or void phase. The void phase can have either air or water or a combination of these two components. There are several useful relationships based on the *volumes* of the various phases present in a given soil that can be defined:

Voids Ratio

Voids Ratio, traditionally denoted as e, is defined as the volume of voids divided by the volume of solids:

$$\text{Voids Ratio} = e = V_V / V_S \qquad (2.2)$$

where:
  V_V = Volume of voids
  V_S = Volume of solids

Voids Ratio is always expressed as a decimal, usually to three decimal places. Loose soils have a high Voids Ratio; compact or dense soils have a low Voids Ratio. Typical ranges are from about 0.300 to 1.200 for different soils, although highly organic soils can have a Voids Ratio of 4.000 or higher.

Porosity

Porosity, traditionally denoted as n, is a term generally used more by geologists than engineers and is defined as the volume of voids divided by the total volume, expressed as a percentage:

$$\text{Porosity} = n = \left(V_V / V_T \right) \times 100\% \qquad (2.3)$$

where:

V_V = Volume of voids

V_T = Total volume

Porosity can only vary from 0%, for a solid mass of material (like a solid steel block), to 100% (for air), but the practical range for most soils is somewhere between about 10%–60%. Using Equations 2.2 and 2.3, we may develop an identity between Voids Ratio and Porosity:

$$n = e / (1 + e) \times 100\% \tag{2.4}$$

Saturation

Saturation, like Water Content, helps describe how much water is present in the soil, but in this case expressed in terms of volumes rather than mass. Saturation is usually denoted as S and is expressed as a percentage. Saturation describes what percentage of the void space (i.e., the volume of the non-solid phase) is filled with water. Saturation is defined as:

$$\text{Saturation} = S = (V_W / V_V) \times 100\% \tag{2.5}$$

where:

V_W = Volume of water

V_V = Volume of voids

For a completely oven dry soil with W = 0%, S = 0%. For a soil in which all of the void space is completely filled with water, S = 100%. This means that the entire range of Saturation is theoretically from 0% to 100%, but a more practical range of natural and compacted soils in the field is probably somewhere between about 20% and 100%, because a soil in the field is never completely dry, as previously described. Generally, it can be assumed for practical purposes that soil samples obtained from below the ground water table are saturated and S = 100%. If S = 100%, we say that the soil is "saturated"; if S < 100%, we say that the soil is "unsaturated".

Phase relationships combining mass and volume

There are several very useful and routinely used relationships that combine the various mass and volume quantities shown in Figure 2.1 that should also be considered and are typically calculated for both natural and compacted soils.

Wet Density

Wet Density (sometimes called *Total Density*) is defined as the total mass (solids + water) divided by the total volume and has units of mass per unit volume, such as g/cm^3 or Mg/m^3 or lbs/ft^3:

$$\text{Wet Density} = \text{Total Density} = \rho_{WET} = M_T/V_T \qquad (2.6)$$

where:
M_T = Total mass
V_T = Total volume

Dry Density

Dry density is defined as the mass of only the solids divided by the total volume and has units of mass per unit volume, such as g/cm³ or Mg/m³ or lbs/ft³:

$$\text{Dry Density} = \rho_{DRY} = \rho_D = M_S/V_T \qquad (2.7)$$

where:
M_S = Mass of solids
V_T = Total volume

It is very useful to be able to convert from Wet Density to Dry Density and vice versa. This can easily be done using the calculated Water Content:

$$\text{Dry Density} = \text{Wet Density}\big/\big[(1 + W/100)\big] \quad \text{or} \quad \rho_{dDRY} = \rho_{WET}\big/\big[(1 + W/100)\big]$$
$$(2.8a)$$

$$\text{Wet Density} = \text{Dry Density} \times \big[(1 + W/100)\big] \quad \text{or} \quad \rho_{WET} = \rho_{DRY} \times \big[(1 + W/100)\big]$$
$$(2.8b)$$

Saturated Density

Saturated Density, ρ_{SAT}, is the special case of Wet Density when $S = 100\%$ and is determined using Equation 2.6. Saturated Density is needed in order to accurately calculate total stress in a soil profile that includes soils below the water table. Figure 2.2 shows the variation in Wet (Total) Density for a number of different undisturbed fine-grained soils determined for a large number of samples taken from below the water table where Saturation may be taken as 100%. The solid line represents the theoretical variation in Wet Density with Water Content for a value of $G = 2.70$ and $S = 100\%$. The variation about the solid line is simply the result of small variations in mass density (Specific Gravity) of the solids, which depends on mineral composition.

Buoyant Density

Buoyant Density is the Saturated Density, ρ_{SAT}, adjusted for the unit density of water, for soils that occur below the water table:

$$\text{Buoyant Density} = \rho_{BOUY} = \rho_{SAT} - \rho_{WATER} \qquad (2.9)$$

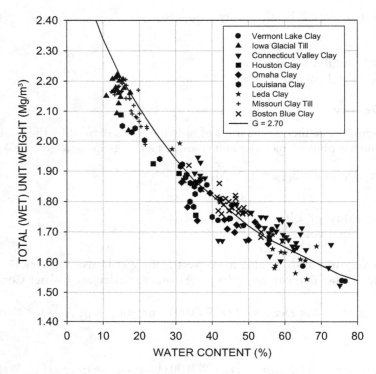

Figure 2.2 Measured variation in Saturated Density as a function of Water Content. (Reference line G = 2.70)

DETERMINING TOTAL UNIT WEIGHT ON TRIMMED SPECIMENS

Standard

ASTM D7263 *Standard Test Methods for Laboratory Determination of Density (Unit Weight) of Soil Specimens*

Equipment

Moisture tins
Ceramic bowls
Spatula or other trimming tools
Convection drying oven capable of maintaining $110° \pm 5°C$
Digital vernier calipers
Electronic scale with a minimum capacity of 500 gm and a resolution of 0.01 gm
Desiccator

Procedure

1. Trim a specimen from a core sample or other type of sample by using a spatula or knife to square off the ends. Trim the specimen so that the height is about 50 mm. Using the digital vernier calipers, measure the height (H) and diameter (D) of the specimen to the nearest 0.1 mm. (Note: If a cubical specimen is being used, trim off all sides and measure the height, width and length [L] of the specimen.) Obtain at least three measurements of each dimension and record all three measurements. Use the average of each measurement and calculate the total volume, V_T, of the specimen.

$$V_T = \pi r^2 H \text{ (for cylindrical specimens)} \tag{2.10a}$$

$$V_T = H \times W \times L \text{ (for cubical specimens)} \tag{2.10b}$$

2. To determine the total mass of the specimen, obtain the mass of a clean dry bowl (M_{BOWL}) to the nearest 0.01 gram using an electronic scale and then place the specimen in the bowl. Now determine the total mass ($M_{BOWL} + M_{WET}$) of the bowl and specimen to the nearest 0.01. The difference between the mass of the wet specimen + bowl and the mass of the bowl (M_{BOWL}) is the total mass of the specimen, M_{WET}.

3. Because all soils in the field contain some water in between the solid mineral particles, i.e., in the void space, it is necessary to drive off all of the water by drying the soil in an oven so that only the solid mineral material remains. To determine the Water Content, place the bowl with the specimen in a convection oven and dry at 110°C until the mass is constant. After drying, remove the bowl from the oven and place it in a desiccator to cool. After cooling, immediately obtain the mass of the bowl and dry soil ($M_{DRY} + M_{BOWL}$). (Note: Don't let the soil sit around in the lab before determining the mass, as it will draw in water from the atmosphere.)

4. Subtract the mass of the bowl from the dry mass of the soil + bowl to obtain the dry mass of soil, M_{DRY}. (Note: If the trimmed specimen is to be used for other testing, then it should not be oven dried and the soil trimmings may be used to determine the Water Content. In this case, place the trimmings in a bowl of known weight and oven dry the trimmings in a convection oven at 110°C until the mass is constant. After drying, obtain the dry mass of the trimmings and then calculate the Water Content.) (See Figures 2.3 and 2.4.)

Calculations

Using the calculated total volume of the specimen, V_T, and the measured values of total mass, M_{WET}, and dry mass, M_{DRY}, (or Water Content), the

Figure 2.3 Measuring the diameter of a trimmed specimen.

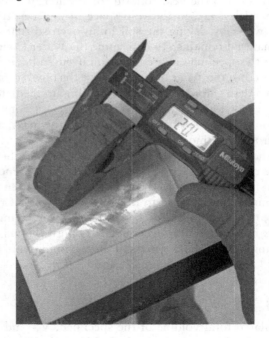

Figure 2.4 Measuring the height of a trimmed specimen.

following quantities may be determined using the phase equations presented at the beginning of this chapter:

Wet (Total) Unit Weight $(\rho_{\text{TOTAL}}$ or $\rho_{\text{WET}}) = M_{\text{WET}}/V_T$

Dry Unit Weight $(\rho_{\text{DRY}}) = M_{\text{DRY}}/V_T$

Water Content $(W) = [(M_{\text{WET}} - M_{\text{DRY}})/M_{\text{DRY}}] \times 100\%$

Voids Ratio $(e) = V_V/V_S$

$(V_S = M_{\text{DRY}}/\rho_S)$

$(V_V = V_T - V_S)$

(Note: Use measured value of ρ_S if it is available; if not, assume $\rho_S = 2.65$ gm/cm^3)

Porosity $(n) = V_V/V_T \times 100\%$

Saturation $(S) = V_W/V_V \times 100\%$

$(V_W = M_{\text{WATER}}/\rho_{\text{WATER}}) = (M_{\text{WET}} - M_{\text{DRY}})/\rho_{\text{WATER}}$

(Note: $\rho_{\text{WATER}} = 1$ gm/cm$^3 = 1$ Mg/m$^3 = 62.4$ lbs/ ft^3)

Example

An undisturbed cylindrical specimen of Boston Blue Clay obtained from a thin-wall Shelby tube sample was trimmed in the lab. The following measurements were obtained:

Diameter = 72.3, 72.3, 72.4 mm (average = 72.33 mm); (average radius = 36.15 mm)

Length = 35.7, 36.3, 35.7 mm (average = 35.9 mm)

Total Volume = $V_T = (\pi)(36.15 \text{ mm})^2(35.9 \text{ mm}) = 147387.6 \text{ mm}^3 = 147.39$ cm^3

Mass of the bowl = 37.11 gm

Mass of the bowl + wet specimen = 305.77 gm

Therefore, the mass of the wet soil = 305.77 gm – 37.11 gm = 268.66 gm

After drying in the oven for 24 h:

Mass of the bowl + dry soil = 226.21 gm

Therefore, the mass of the dry soil = 226.21 gm – 37.11 gm = 189.10 gm.

Determine the various phase relationship quantities.

Wet (Total) Unit Weight

$(\rho_{\text{TOTAL}}$ or $\rho_{\text{WET}}) = 268.66 \text{ gm} / 147.39 \text{ cm}^3 = M_{\text{WET}}/V_T$

$$= 1.82 \text{ gm/cm}^3 = 1820 \text{ kg/m}^3 = 113.6 \text{ lbs./ft.}^3$$

Dry Unit Weight

$\rho_{\text{DRY}} = M_{\text{DRY}}/V_T = 189.10 \text{ gm}/147.39 \text{ cm}^3 = 1.28 \text{ gm/cm}^3$

$$= 1280 \text{ kg/m}^3 = 79.9 \text{ lbs./ft}^3$$

Water Content

$$W = \left[(M_{WATER} - M_{DRY})/M_{DRY} \right] \times 100\%$$

$$= \left[(268.66\,gm - 189.10\,gm) \right]/189.10\,gm \times 100\% = 42.1\%$$

Voids Ratio

$$e = V_V/V_S$$

$$V_S = M_{DRY}/\rho_S = 189.10\,gm \,/\, 2.65\,gm/cm^3 = 71.36\,cm^3$$

$$V_V = V_T - V_S = 147.39\,cm^3 - 71.36\,cm^3 = 76.03\,cm^3$$

$$e = 76.03\,cm^3/71.36\,cm^3 = 1.065$$

(note: A value of G=2.65 (ρ_s=2.65 gm/cm³) was assumed for this soil.)

Porosity

$$n = V_V \,/\, V_T \times 100\% \; = 76.03\,cm^3/147.39\,cm^3 \times 100\% = 51.6\%$$

Saturation

$$S = V_W/V_V \times 100\%$$

$$V_W = M_{WATER}/\rho_{WATER} = (268.66\,gm - 189.10\,gm)/1\,gm/cm^3 = 79.56\,cm^3$$

$$S = \left(79.56\,cm^3/76.03\,cm^3 \right) \times 100\% = 104.6\%$$

How can the calculated Saturation be greater than 100%? There are two possible explanations: 1) the assumed value of Specific Gravity of the solids may be off a little (i.e., instead of the assumed value of G=2.65, it may be 2.70); 2) there were some small errors in determining volume of mass.

Example

A soil sample was collected from below the water table. A wet sample with a total mass of 320.6 gm was dried in a convection oven for 24 h. After 24 h, the dry mass was determined to be 231.3 gm. Determine the Water Content, Wet Density and Dry Density.

Water Content

$$M_{WATER} = M_{WET} - M_{DRY} = 320.6\,gm - 213.3\,gm = 89.3\,gm$$

$$W = \left(M_{WATER}/M_{DRY}\right) \times 100\% = (89.3\,gm/231.3\,gm) \times 100\% = 38.6\%$$

Since the sample was obtained below the water table, it is very reasonable to assume that the sample is saturated, i.e., S = 100%. A simple trick that is used in this case and which does not affect the outcome of the calculations is to first assume a unit volume of solids; for example, assume that $V_S = 1$ cm³. Again, using an assumed value of Specific Gravity of 2.70 ($\rho_s = 2.70$ gm/cm³), it is possible to calculate the various quantities.

$$M_{DRY} = V_S \times \rho_S = 1\,cm^3 \times 2.70\,gm/cm^3 = 2.70\,gm$$

(Note: $\rho_{WATER} = 1.0$ gm/ cm³ = 1000 kg/m³ = 62.4 lbs/ft³)

Since W = 38.6% and W = $(M_{WATER}/M_{DRY}) \times 100\%$, $M_{WATER} = M_{DRY} \times$ (W/100) = (2.70 gm × 0.386 = 1.04 gm

$$V_W = M_{WATER}/\rho_{WATERr} = 10.4\,gm \,/\, 1.0\,gm \,/\, cm^3 = 1.04\,cm^3$$

Wet (Total) Unit Weight

$$\rho_{WET} = M_{WET}/V_T = \left(2.70\,gm + 1.04\,gm\right)\Big/\left(1\,cm^3 + 1.04\,cm^3\right)$$

$$= 1.83\,gm/cm^3 = 1830\,kg/m^3 = 114.2\,lbs./ft^3$$

Dry Unit Weight

$$\rho_{DRY} = M_{DRY}/V_T = \left(2.70\,gm\right)\Big/\left(1.0\,cm^3 + 1.04\,cm^3\right)$$

$$= 1.32\,gm/cm^3 = 1320\,kg/m^3 = 82.4\,lbs./ft^3$$

or

$$\rho_{DRY} = \left(\rho_{WET}\right)\Big/(1 + W/100) = \left(114.2\,lbs./ft^3\right)\Big/(1 + 0.386) = 82.4\,lbs./ft^3$$

It is now possible to calculate Voids Ratio and Porosity if needed using Equations 2.2 and 2.3.

Discussion

Determination of Water Content, Unit Weight, etc. forms the basis of nearly all laboratory soil testing and is used to describe soil samples that are used on many other tests. Errors generally occur in improperly obtaining or recording measurements of specimen dimensions or mass or both. As noted, in the first example given above, the calculated value of Saturation was greater than 100%. Since this is physically impossible, the most likely explanation is that the assumed value for the density of solids (Specific Gravity) is slightly off, but this type of error is usually not critical.

However, remember that the other quantities, such as Water Content and Unit Weight, are based on *measured* values and therefore are most likely correct unless a math error has occurred. Even with the calculated Saturation value greater than 100%, the quantities are considered reliable enough for most routine work. Table 2.1 gives some typical values of measured Wet Unit Weight, Water Content and calculated Dry Unit Weight for a variety of natural fine-grained soils.

At any given site, there will usually be natural variation in Water Content and Dry Unit Weight. For example, Figure 2.5 shows the variation in Water Content and Dry Unit Weight at a site in northern New York.

Table 2.1 Typical Values of Wet Unit Weight, Water Content and Dry Density for a Variety of Soils

Soil	Wet Unit Weight lbs/ft³ (Mg/m³)	Water Content (%)	Dry Unit Weight lbs/ft³ (Mg/m³)
Clayey Loess	116.8 (1.87)	30.0	89.9 (1.44)
Silty Loess	111.4 (1.79)	15.9	96.1 (1.54)
Connecticut Valley Varved Clay	106.4 (1.71)	56.4	68.0 (1.09)
Iowa Glacial Till	137.1 (2.20)	15.5	119.2 (1.91)
London Clay	127.4 (2.04)	23.5	103.2 (1.65)
Boston Blue Clay	115.8 (1.86)	38.9	83.4 (1.34)
Mississippi Alluvial Clay	110.4 (1.77)	43.9	76.7 (1.23)
Nebraska Alluvial Clay	108.9 (1.75)	48.6	73.3 (1.17)
Leda Clay	102.2 (1.64)	62.2	63.0 (1.01)
Louisiana Alluvial Clay	119.9 (1.92)	31.8	91.0 (1.46)
Houston Beaumont Clay	118.3 (1.90)	23.0	96.2 (1.54)

DETERMINING TOTAL UNIT WEIGHT USING THE ELEY VOLUMETER

Another quick and easy method that can be used to quickly obtain a measurement of the Unit Weight of soils in the field, such as from a road cut, excavation or trench, is the Eley Volumeter. The volumeter is a small calibrated stainless-steel cylinder, shown in Figure 2.6. The cylinder has an internal plunger and a calibrated threaded stem that can be read to the nearest 0.05 cm³. When the stem is retracted, the cylinder has a capacity of 30 cm³. The end of the cylinder has a sharpened cutting edge that is pushed slowly into the soil as shown in Figure 2.6.

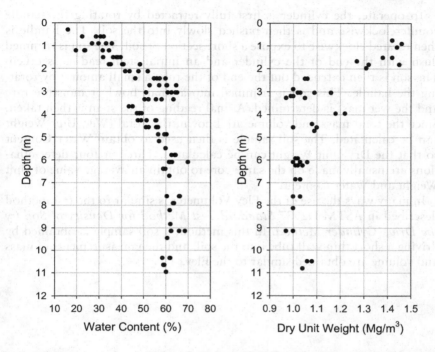

Figure 2.5 Variation in natural Water Content and Dry Unit Weight at a site of Champlain Sea Clay in northern New York.

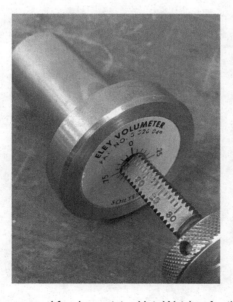

Figure 2.6 Eley Volumeter used for determining Unit Weight of soils.

To operate, the cylinder is first fully retracted by rotating the handle counterclockwise and is then pushed slowly into the soil. The handle is then rotated clockwise to expose a short section of soil. The soil is trimmed flush with the end of the cylinder and an initial stem reading is taken. The soil is then extruded out the end of the tube a small amount by rotating the handle. This soil is trimmed into a small bowl or moisture tare and the wet mass is determined. A final reading of the stem is then taken. Since the total mass and volume are known, the Total (Wet) Unit Weight can be calculated. The soil sample is then used to obtain Water Content so that the Dry Unit Weight can be calculated. Three or four determinations are usually made on the same core to obtain an average value of Unit Weight and Water Content.

In many ways the use of the Eley Volumeter is similar to the test method described in ASTM D2937 *Standard Test Method for Density of Soil by the Drive-Cylinder Method*. In this method, a soil sample is obtained by driving a short thin wall tube into the soil, and then measurements of mass and volume are obtained similar to the Eley.

Chapter 3

Density (Unit Weight) of Solids (Specific Gravity)

INTRODUCTION

The density or Unit Weight of the solids, ρ_s, is an important property of soils. As described in Chapter 2, the Unit Weight of the solids is needed to calculate the various phase components of a soil mass, such as Voids Ratio, Porosity and Saturation. The Unit Weight of solids is an important quantity for these calculations and should be measured whenever possible. The Unit Weight of solids is also often called *Specific Gravity*, which can be thought of as the ratio of the Unit Weight of any material to the Unit Weight of water.

$$G_S = \rho_S / \rho_{WATER} \tag{3.1}$$

where:
G_S = Specific Gravity
ρ_s = Unit Weight of soil
ρ_{WATER} = Unit Weight of water

Water has a Unit Weight of 1.0 gm/cm^3 or 1000 kg/m^3 or 62.4 lbs/ft^3, and water has a Specific Gravity of 1.0. The Unit Weight of soil solids depends largely on the mineralogic composition, which varies somewhat with geologic origin. The test is best performed on oven-dry soil, and the procedure presented in this chapter is for oven-dry soil. The water pycnometer method described in this chapter is intended for soil passing the No. 4 (0.425 mm) sieve. For larger size soils, such as sands and gravels, ASTD C128 *Standard Test Method for Density, Relative Density (Specific Gravity), and Absorption of Fine Aggregates* should be used (Figure 3.1).

STANDARD

ASTM D854 *Specific Gravity of Soil Solids by Water Pycnometer*

DOI: 10.1201/9781003263289-4

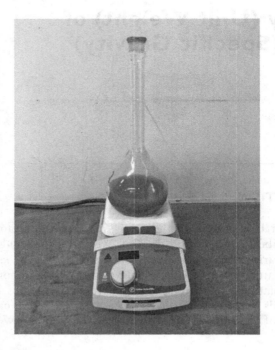

Figure 3.1 Pycnometer on a hot plate.

EQUIPMENT

500 mL pycnometer and cap
10 mL pipette
De-aired water
Electronic scale with a minimum capacity of 1000 gm and a resolution
   of 0.01 gm
Plastic wash bottle
Thermometer with a range of 0–50°C and a resolution of 0.5°C
Desiccator
Small glass funnel
Convection oven capable of maintaining 110° ± 5°C
Hot plate or vacuum pump

PREPARATION

1. For fine-grained soil, initially air-dry a representative soil specimen
   and use a mortar and rubber-tipped pestle to crush enough sample
   to pass a No. 40 sieve. Place the specimen in a convection oven for a
   minimum of 24 h to obtain an oven-dry specimen.

Table 3.1 Minimum Recommended Oven Dry Mass of Sample for Determining Specific Gravity (after ASTM D854)

Soil type	Minimum oven-dry mass for 250 mL pycnometer	Minimum oven-dry mass for 500 mL pycnometer
SP, SP-SM	60 ± 10 gm	100 ± 10 gm
SP-SC, SM, SC	45 ± 10 gm	75 ± 10 gm
Silt or clay	35 ± 5 gm	50 ± 10 gm

2. To help reduce potential errors in this test, a minimum oven-dry mass of soil should be used. ASTM D854 gives recommended mass based on soil type, as given in Table 3.1.

CALIBRATION OF PYCNOMETER

1. Select a clean, dry 250 mL or 500 mL glass pycnometer with a calibration mark on the neck.
2. Carefully fill the pycnometer to the calibration mark on the neck with distilled water. Make sure that there are no water drops on the outside of the pycnometer. Determine the mass of the filled pycnometer and record this mass as M_a.
3. Use a thermometer and determine the temperature of the distilled water in the pycnometer to the nearest $0.5°C$ and record this value as T_b.
4. ASTM D854 requires that calibration of the empty dry pycnometer and pycnometer + water be performed five times.

PROCEDURE

1. Obtain the equivalent oven-dry mass of the soil specimen to the nearest 0.01 gm using an electronic scale. (Note: use moist or air-dried soil and obtain the Water Content as described in Chapter 1.) Record this value as M_s.
2. Transfer the soil to the pycnometer using a small glass funnel. (Note: Be sure to wash down any soil adhering to the funnel and the neck of the pycnometer into the pycnometer using a small wash bottle to make sure that all of the soil falls to the bottom of the pycnometer.)
3. Fill the pycnometer about half way with distilled water.
4. Remove the trapped air from the soil/water mixture by one of the following methods:
   a.) Boil the mixture gently using a hot plate or a Bunsen burner for at least 10 minutes while carefully agitating the pycnometer to help trapped air escape. Then let the pycnometer cool to room temperature; or

b.) Place a stopper on the top of the pycnometer and attach the stopper to a vacuum pump with a short piece of tubing and run the pump at no more than 100 mm Hg (2.5 psi) for 30 minutes. While the vacuum is being applied, gently agitate the pycnometer to help air escape.

(Note: De-airing of the pycnometer is an extremely important step. Most errors in this test are the result of trapped air. Oven-dry samples of some clays may require 2–4 h to remove all trapped air. During de-airing, watch the soil–water mixture to see if any more bubbles occur.)

5. After de-airing and cooling, fill the pycnometer to the mark with distilled water using the plastic wash bottle. Determine the total mass of the pycnometer with soil and water and record this mass as M_b.

CALCULATIONS

Calculate the density of solids in gm/cm^3 to the nearest 0.01 gm as:

$$\rho_S \left(\text{at } T_b \right) = M_s / \left[M_s - \left(M_a - M_b \right) \right] \tag{3.2}$$

where:
ρ_S = density of solids (gm/cm^3)
M_s = oven-dry mass of soil (gm)
M_a = mass of pycnometer filled to mark with only distilled water at temperature T_b from pycnometer calibration
M_b = mass of pycnometer filled with soil and distilled water to mark at temperature T_b

The value of ρ_s should be reported to a standard laboratory temperature of 20°C as:

$$\rho_S \text{ at } 20°C = K \left(\rho_s \text{ at } T_b \right) \tag{3.3}$$

where:
K = a correction factor based on the density of water from Table 3.2.

EXAMPLE

The following results were obtained from a sample of sandy silty clay glacial till from Illinois:

Pycnometer No. 12
Mass of oven-dry soil = 52.20 gm

Table 3.2 Correction Factors for Temperature

Temperature (°C)	Density of water (gm/cm³)	Correction factor K
16.0	0.99897	1.0007
16.5	0.99889	1.0007
17.0	0.99880	1.0006
17.5	0.99871	1.0005
18.0	0.99862	1.0004
18.5	0.99853	1.0003
19.0	0.99843	1.0002
19.5	0.99833	1.0001
20.0	**0.99823**	**1.0000**
20.5	0.99812	0.9999
21.0	0.99802	0.9998
21.5	0.99791	0.9997
22.0	0.99780	0.9996
22.5	0.99768	0.9995
23.0	0.99757	0.9993
23.5	0.99745	0.9992
24.0	0.99732	0.9991

Mass of pycnometer + soil + water = 706.80 gm.
Mass of pycnometer + water = 674.28 gm
Temperature = 23.5°C
ρ_S = 52.20gm/[(52.20 gm − (706.80gm − 674.28 gm)] = 2.65
From Table 3.2, K = 0.9992
ρ_S (@ 20°C) = 0.9992 (2.65) = 2.65

DISCUSSION

In this test, small errors in the measurement of the mass of the oven-dried soil and the mass of the pycnometer as well as filling of the pycnometer to the same exact mark and the temperature can significantly influence the results. The most common problems develop from insufficient de-airing of the pycnometer and insufficient removal of trapped air in the soil. Another potential source of error in this test is not adequately allowing for thermal equilibrium of the water. It is recommended that a reservoir of de-aired water (3–5 L) is used for calibration and testing should be allowed to come to thermal equilibrium in a constant temperature chamber before testing.

Air-drying specimens of soil before performing the test for Specific Gravity will not normally produce significant errors in the test results; however, oven drying should be avoided, especially for soils that contain organic materials. For example, Lambe (1951) showed that oven-drying soils prior to performing the test for Specific Gravity had little to no effect on the values obtained inorganic clays but could substantially increase the measured value (from about 1.80 to 2.61) for some fine-grained soils with organics.

The laboratory test for determining Specific Gravity can be time consuming and can often be difficult to conduct, especially if great care is not taken to obtain volumes and masses. An alternative method of determining Specific Gravity of soils using a gas pycnometer is described in ASTM D5550 *Standard Test Method for Specific Gravity of Soil Solids by Gas Pycnometer*. This method may be attractive for some laboratories.

Unit Weight of water

As previously noted, the unit density (sometimes referred to as Specific Gravity) of water is typically taken as a constant: $\rho_{WATER} = 1.0$ g/cm^3 = 1000 kg/m^3 = 62.4 lbs/ft^3 (i.e., $G_{WATER} = 1.0$)

Unit Weight of solids

The Unit Weight of the solid phase of soils depends somewhat on the mineralogic composition and traditionally was called Specific Gravity (G) to indicate its comparison to the reference value of water. For most practical purposes, the actual value of unit density does not vary that much for a wide range of soils, and typically, for routine work, it can be taken as: $\rho_{SOLIDS} = 2.65$ to 2.75 gm/cm^3 = 2650 to 2750 kg/m^3 = 165.4 to 171.6 lbs/ft^3; $G_S = 2.65$ to 2.75). For most sands, a reasonable value of ρ_S is 2.65 gm/cm^3, which is the density of quartz (SiO$_2$). The value of ρ_S for clays and other fine-grained or mixed-grain soils is more variable because of differences

Table 3.3 Typical Measured Values of Specific Gravity for a Variety of Soils

Soil	Specific Gravity	Unit Weight of solids lbs/ft^3 (gm/cm^3)
Nebraska Loess	2.83	176.6 (2.83)
Georgia Residual Clay	2.68	167.2 (2.68)
Iowa Glacial Till	2.73	170.3 (2.73)
Connecticut Valley Varved Clay	2.81	175.3 (2.81)
Leda Clay	2.64	164.7 (2.64)
Boston Blue Clay	2.78	173.5 (2.78)
Oklahoma Clay	2.80	174.7 (2.80)
Kaolinite KGa-1b	2.68	167.2 (2.68)
Na-Montmorillonite SWy-2	2.75	171.6 (2.75)
Ca-Montmorillonite STx-1	2.70	168.5 (2.70)

in grain-size distribution, composition, Organic Content and mineralogy. Table 3.3 gives measured values of ρ_S for common pure clay minerals and for a range of some natural soils.

REFERENCE

Lambe, T.W., 1951. *Soil Testing for Engineers*. John Wiley & Sons: New York.

Chapter 4

Grain-Size Distribution

BACKGROUND

One of the most fundamental measurements that can be made on the solid phase of soil is used to describe the assemblage of individual grains or particles that make up that soil. This is done by determining the mass of soil grains of different sizes by separating the grains. This procedure is referred to as the *grain-size analysis* or *particle-size analysis* and results in a grain-size distribution curve for the soil. The grain-size analysis is performed in two parts: 1) the sieve analysis, and 2) the hydrometer or pipette analysis.

The sieve analysis is used for the coarse fraction and separates particles into different sizes by passing the soil through a series of screens or sieves. The sieves are stacked with the largest size sieve opening at the top and the smallest size sieve opening at the bottom. A pan is placed below the smallest sieve on the bottom to catch any soil that falls through the last sieve. The stack of sieves is usually called a "nest" of sieves. Dry soil is poured into the top sieve, the lid is placed on top and the sieves are placed on a mechanical shaker so that particles fall by gravity until they are held on a screen with an opening smaller than the equivalent diameter of the soil particle. After shaking, the sieves are taken apart and weighed so that the mass of soil held or retained on that sieve is determined. The sieve analysis works best for coarse-grained soils, such as gravels and sand, which have particles that are large enough to be seen by the naked eye and are easily separated.

Since some of the soil particles are so small and it is not practical to manufacture wire sieves with small enough openings, the sieve analysis can't be used to determine the size of very small soil particles, i.e., less than about 0.075 mm (No. 200 sieve). For these particles, we need to use a different procedure, either the hydrometer analysis or the pipette analysis, which separates particles of different sizes based on how fast they fall through a suspension with distilled water. Different common terms are used by engineers to describe different size soil particles, as shown in Table 4.1. Some systems of soil identification use 0.0063 mm to define sand and distinguish the fines content (silt + clay). This size is also often used by geologists and agronomists.

DOI: 10.1201/9781003263289-5

Table 4.1 Common Terms to Describe
Different Particle Sizes

Term	Size (mm)
Clay	< 0.002
Silt	0.002–0.075
Fine sand	0.075–0.42
Medium sand	0.42–2.0
Coarse sand	2.0–4.75
Gravel	4.75–75

COARSE FRACTION – SIEVE ANALYSIS

Introduction

The distribution of particle sizes is commonly used to describe soils and can be of some use in obtaining engineering properties in conjunction with empirical relationships. The grain-size distribution of coarse-grained soils is determined by using a series of sieves ranging in aperture size. The procedure described below is for dry sieving of soil samples and generally is best suited to coarse-grained soils with > 50% retained on the No. 200 (0.075 mm) sieve.

Standard

ASTM D6913 *Standard Test Methods for Particle-Size Distribution (Gradation) of Soils Using Sieve Analysis (Supersedes* ASTM D422 *Standard Test Method for Particle Size Analysis of Soils)*

Equipment

Brass or stainless-steel sieves (3/8", No.4, No.10, No.20, No.40, No.60, No.100, No.200, lid, pan)
Electronic scale with a minimum capacity of 1000 gm and a resolution of 0.1 gm
Mechanical shaker
Sieve cleaning brush
Convection oven capable of maintaining $110° \pm 5°C$

Procedure

1. Select a set of sieves in accordance with ASTM D6913. (Note: The sieves selected for determining grain-size distribution may vary depending on the soil. Using more sieves helps define the grain-size distribution curve more accurately, especially for coarse-grained soils.)

2. Clean and dry each sieve. Table 4.2 gives size openings of common standard sieves.
3. Determine the mass of each clean sieve to the nearest 0.1 gm (M_{SIEVE}).
4. Place the sieves in order in the mechanical shaker with the largest sieve on the top and the finest sieve on the bottom.
5. Select a representative sample of oven-dried soil, breaking up any obvious soil clumps. Determine the total mass to the nearest 0.1 g (M_{SOIL}). Typical oven-dry masses of soil suggested by ASTM D6913 are given in Table 4.3. (Note: It may be necessary to initially break up the soil by hand or use a ceramic mortar and rubber-tipped pestle to gently break up the soil.)
6. Pour the soil into the top sieve and place a lid on the top of the sieve nest.
7. Clamp the nest of sieves in the shaker.
8. Turn on the mechanical sieve shaker and let it run for approximately 10 minutes.

Table 4.2 Size Openings of Common Sieves

Soil type	Sieve no.	Screen opening (mm)
Coarse-grained (sieve analysis)	¾	19.05
	3/8	9.52
	4	4.75
	6	3.35
	8	2.36
	10	2.00
	12	1.68
	16	1.18
	20	0.85
	30	0.60
	40	0.425
	50	0.300
	60	0.250
	80	0/180
	100	0.150
	140	0.106
	200	0.075

Table 4.3 Recommended Soil Mass for Use in Sieve Analysis

Nominal diameter of largest particle	Approximate minimum mass of soil
< 3/8 in	500 grams
< ¾ in	1000 grams

9. After shaking, remove and determine the mass of each sieve, starting at the top of the nest and progressing down. Be careful not to spill any soil when removing the sieves from the nest. Record the combined mass of each sieve plus soil: M_{TOTAL} (Figure 4.1).

Calculations

Calculate the mass of soil retained on each sieve as:

$$M_{SOIL} = M_{TOTAL} - M_{SIEVE}$$

Calculate the percent of soil finer for each grain size as:

$$\%Finer = 1 \ (Mass\ Retained/Total\ Mass\ of\ Soil) \times 100\%$$

Prepare a semi-log plot of grain diameter (x-axis log scale) versus % finer (y-axis).
　　Calculate the following parameters:

Mean grain diameter

$$D_{50} = grain\ size(diameter)for\ which\ 50\%\ of\ soil\ is\ finer \tag{4.1}$$

Uniformity Coefficient

$$C_U = D_{60}/D_{10} \tag{4.2}$$

Figure 4.1 No. 4 (left) and No. 40 (right) standard sieves.

D_{60} = grain size (diameter) for which 60% of soil is finer
D_{10} = grain size (diameter) for which 10% of soil is finer

Coefficient of Curvature

$$C_C = (D_{30})^2 / (D_{10} \times D_{60})$$ (4.3)

D_{30} = grain size (diameter) for which 30% of soil is finer

Percentage fines

Fines = % passing No. 200 sieve (4.4)

Example

The results in Table 4.4 were obtained for a sample of sand from Plymouth, Ma.

Initial oven-dry mass of soil = 371.3 gm.

The results from the test are presented as a grain-size distribution curve in Figure 4.2. From the grain-size distribution curve shown in Figure 4.2, the following parameters were determined:

D_{10} = 0.16 mm
D_{30} = 0.36 mm
D_{50} = 0.64 mm
D_{60} = 0.77 mm
Mean grain diameter = D_{50} = 0.64 mm

Table 4.4 Sample of Sand

Sieve no.	Mass of sieve (gm)	Mass of sieve + soil (gm)	Mass of soil retained (gm)	Mass passing (gm)	% finer
¾"	556.3	556.3	0.0	371.3	100.0
3/8"	535.5	549.7	14.2	357.1	96.2
No. 4	501.2	526.5	25.3	331.8	89.4
No. 10	471.2	501.6	30.4	301.4	81.2
No. 20	576.5	643.3	66.8	234.6	63.2
No. 40	339.8	446.0	106.2	128.4	34.6
No. 60	323.8	390.6	66.8	61.6	16.6
No. 100	372.7	400.2	27.5	34.1	9.2
No. 200	340.3	353.3	13.0	21.1	5.7
Pan	346.6	367.7	21.1	0.0	

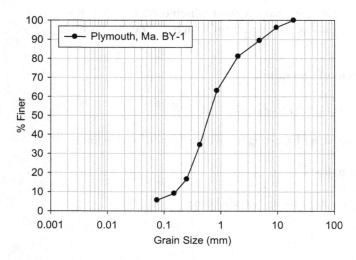

Figure 4.2 Grain-size distribution curve of Plymouth Sand.

Uniformity Coefficient = C_U = D_{60}/D_{10} = 0.77mm/0.16 mm = 4.81
Coefficient of Curvature = C_C = $(D_{30})^2/(D_{10} \times D_{60})$ = $(0.36)^2/(0.16 \times 0.77)$
   = 1.05
% fines = % passing No. 200 sieve = 5.7%

Discussion

The cumulative grain-size distribution curve can be used to provide several important quantitative measures of both the shape and position of the curve. These parameters provide a useful means of comparing different coarse-grained soils and should be reported. Laboratory sieve analyses are relatively inexpensive and quick to perform and can reveal a lot about the composition and variability of the soil.

The uniformity Coefficient is often used to create terms to describe the general nature of the overall grain-size distribution for coarse-grained soils. The term *well-graded* indicates that the soil is composed of individual particles over a wide range of sizes in the grain-size curve, while the term *uniform* indicates that most of the particles fall within a narrow range. Table 4.5 gives recommended guidelines for identifying well-graded and uniform (poorly-graded) sands and gravels based on the Unified Soil Classification System (USCS) using results of the sieve analysis.

The gradation of coarse-grained soils can have a profound influence on engineering behavior. Well-graded soils are easier to compact and have higher strength and lower compressibility as compared to uniform soils. The wide range of particle sizes of a well-graded soil allows for tighter packing as the smaller particles fit into the void space between larger particles

Table 4.5 Requirements for Well-Graded and
Uniform Coarse-Grained Soils

Soil	Grading	C_c	C_U
Sand	Well-graded	$C_c > 6$	$C_U = 1-3$
Gravel	Well-graded	$C_c > 4$	$C_U = 1-3$

as compared to soils that have particles of about the same size. Results of the grain-size distribution curve can also be used to estimate minimum and maximum index Unit Weights for sands and gravelly sands, which can be used to calculate minimum and maximum void ratios (e.g., Lutenegger 2019).

FINE FRACTION – HYDROMETER ANALYSIS

As previously noted, it is difficult to determine the distribution of fine particles using the mechanical sieve analysis, simply because wire screens cannot be manufactured with small enough openings. Plus, the fine particles may be held to each other in aggregates that need to be broken up so that the number of each size of individual particle can be determined. Smaller size sieves are available; for example, the No. 325 sieve has an opening of 0.045 mm, the No. 400 sieve has an opening of 0.038 mm and the No. 600 sieve has an opening of 0.020 mm. However, these sieves are rarely used in routine work but might have an application in some research work.

Introduction

The Hydrometer Test is an indirect technique used to determine the grain-size distribution for the fine-grained particle size range. This technique is based on the theory of Stoke's Law for the terminal velocity of spheres falling through a viscous medium. To obtain the velocity of fall of soil particles, a hydrometer is used. This instrument consists of a lead-weighted glass tube that is calibrated to read either the Specific Gravity (151H) or grams/liter (152H) of a soil–water suspension. The suspension is formed by mixing a given quality of soil with water and a small amount of dispersing agent to form a 1000 mL quantity of suspension.

A dispersing agent is usually added to neutralize the particle charges on the smaller soil particles and allow them to separate. Charged particles can form larger particles called *flocs*, which settle out of suspension faster than the true individual particle sizes. During the test, readings are taken at specified times by carefully placing the hydrometer in the soil–water suspension. Details of the procedures and theory associated with the Hydrometer Test can be found in ASTM D7928.

Standard

ASTM D7928 *Standard Test Method for Particle-Size Distribution (Gradation) of Fine-Grained Soils Using the Sedimentation (Hydrometer) Analysis.* (Supersedes ASTM D422 *Standard Test Method for Particle Size Analysis of Soils.*)

Equipment

No. 200 sieve
Electronic scale with a minimum capacity of 500 gm and a resolution of
    0.01 gm
1000 mL hydrometer cylinders
Mixer and mixing cup or Air-Jet Dispersion apparatus
Thermometer 0–100°C
500 mL beakers
Glass stirring rod
Distilled water
Small glass watch glass or plate
25 mL glass beakers
Sodium hexametaphosphate (dispersing agent)
Plastic wash bottles
Rubber stoppers to fit 1000 mL hydrometer cylinders
Stopwatch
152H hydrometer
Convection oven capable of maintaining 110° ± 5°C

Procedure

1. Prepare a reference solution of distilled water and dispersing agent by adding 5.00 gm of sodium hexametaphosphate to 125 mL of distilled water in a 500 mL beaker. Dissolve thoroughly.
2. Obtain the equivalent of approximately 50 gm of oven-dry soil. (Note: Some laboratories do not oven-dry the soil, but instead use an air-dried specimen of soil and determine the Water Content from $M_{WET} = M_{DRY}(1 + W/100)$ using a separate sample to determine the Water Content).
3. Mix the soil to a thick slurry using 125 mL of a 40 gm/liter solution of sodium hexametaphosphate and distilled water as necessary. Stir until the soil is thoroughly wetted. (Note: To maintain better control over the weight of dispersant added to the soil, individual solutions of sodium hexametaphosphate can be prepared for each specimen [i.e., 5 gm per specimen]).
4. Cover the beaker with a glass plate or plastic wrap to prevent any foreign particles from entering the solution and allow the solution to soak for a minimum of 16 hours.

Figure 4.3 1000 mL hydrometer cylinder.

5. Prepare a reference liquid consisting of distilled water and dispersing agent in the same proportions as in the sedimentation test, i.e., 40 gm/L.

6. Transfer the slurry to a malt mixer and add 125 mL of distilled water and mix for 1 minute. Take care to not lose any portion of the sample during transfer. (Note: If using the Air-Jet Dispersion apparatus, transfer the soil solution to a 1000 mL hydrometer cylinder and fill with distilled water to 250 ml. Adjust the air pressure to 172 kPa (25 psi) and disperse for 5 min.)

7. After Dispersion, transfer the slurry to a 1000 mL cylinder and add distilled water as necessary to bring the total volume to 1L as shown in Figure 4.3. (Note: If using the Air-Jet Dispersion apparatus, carefully remove the Air-Jet and use a wash bottle of distilled water to wash all soil particles from the Air-Jet into the cylinder.)

8. Mix the solution for 1 minute by placing a rubber stopper or a hand over the open end of the hydrometer cylinder and then turning the cylinder upside down and back. Set the cylinder down and simultaneously start a timer (t = 0). At about the same time, make sure that the reference solution is also mixed for 1 minute using the same procedure.

9. Carefully place the hydrometer in the suspension and obtain hydrometer readings at 1, 2, 4, 8, 16 min, etc. Take readings for at least

18 h so that the percentage finer than 0.002 mm can be determined. (Note: The hydrometer should be read at the top of the meniscus, and after the 4-minute reading it should be removed from the suspension, rinsed and placed in the reference liquid between readings. Also, the cylinder with the soil suspension should be covered with a glass plate between readings.)

10. Correct the hydrometer readings for temperature, the dispersing agent and the meniscus by taking a reading in the reference solution each time a hydrometer reading is obtained from the sample. Make a separate determination of the meniscus correction using distilled water.

11. At the end of the experiment, gently shake the suspension and then pass the suspension through a No. 200 sieve to separate out the sand fraction. Use a wash bottle to wash the sand from the sieve into a small glass beaker. Place the beaker in a convection oven. After 24 h, remove the beaker and determine the oven-dry mass of the sand fraction.

Calculations

For the Hydrometer Test, the maximum particle diameter still in solution at and/or above the center of volume of the hydrometer is computed as:

$$D = K \times (L/T)^{0.5} \tag{4.5}$$

where:
  D = particle diameter, mm
  L = effective depth, cm (Table 4.2, ASTM D6913; see Appendix II). Use original hydrometer reading corrected only for the meniscus.
  T = elapsed time in minutes
  K = constant (Table 4.3, ASTM D6913; see Appendix II)

For hydrometer 152H, the percent passing is computed as:

$$P(\%) = Ra/M_{DRY} \times 100\% \tag{4.6}$$

where:
  R = corrected hydrometer reading (actual reading–control fluid reading)
  M_{DRY} = oven-dry mass of soil
  a = correction factor (Table 4.1, ASTM D422; see Appendix II)

Example

The Hydrometer Test results in Table 4.6 were obtained from a 50.0 gm oven-dry mass of silty clay soil from Hadley, Ma.

Table 4.6 Hydrometer Test Results

Time (min)	Temp (°C)	Hydrometer reading	Reference solution reading	Diameter (mm)	% finer
1	26.0	43.5	3.5	0.0376	91.7
2	26.0	42.5	3.5	0.0268	89.4
4	26.0	42.0	3.5	0.0190	88.4
8	26.0	41.5	3.5	0.0135	87.1
170	24.5	33.0	4.5	0.0032	65.3
550	24.5	24.5	4.5	0.0019	45.8
1130	23.5	23.0	5.0	0.0014	41.3
1940	24.0	22.0	5.0	0.0010	39.0

Mass of dry soil retained on the No. 200 sieve = 1.3 gm. % sand = (1.3 gm/50.0 gm) × 100% = 2.6%. Individual amounts for each grain size are interpolated from the hydrometer results.

Discussion

For fine-grained soils it is useful to define some specific grain-size ranges to describe the composition. The designations given in Table 4.7 are often used.

In the example above:

Sand = 2.6%
Silt = 51.5%
Coarse silt = 8.6%
Fine Silt = 88.8% − 45.9% = 42.9%
Coarse Silt = 100% − 2.6% − 45.9% = 42.9%
Clay = 45.9%

FINE FRACTION – PIPETTE ANALYSIS

Introduction

An alternative to the Hydrometer Test method for determining the size distribution of the fines fraction of soils is the Pipette Test method, which is commonly used by agronomists. The pipette method is also based on Stoke's Law and assumes that dispersed particles of various shapes and sizes free fall in a liquid according to mass. The pipette method uses a small glass pipette to draw samples from a soil–water suspension after different settling times. The oven-dry mass of the specimens is then determined, and calculations are made to determine the percentage for each corresponding particle size.

Table 4.7 Common Terms Used to Describe Fine-Grained Soils

Term	Size (mm)
Sand	> 0.075
Coarse silt	0.075–0.020 mm
Fine silt	0.020–0.002 mm
Clay	< 0.002 mm
Fine clay (colloids)	< 0.001 mm

Standard

There is currently no ASTM standard test procedure for the Pipette Test method. ISO 17892-4 describes the pipette method. A standard procedure is also given in British Standard BS1377. In this chapter, the procedure described by Kilmer and Alexander (1949) and Walter et al. (1978) and used extensively by the author is described.

Equipment

No. 200 sieve
Electronic analytical scale with a minimum capacity of 100 gm and a resolution of 0.0001 gm
1000 mL hydrometer cylinders
Convection Oven capable of maintaining 110° ± 5°C
Mixer and mixing cup or Air-Jet Dispersion apparatus
Thermometer with a range of 0–50°C with a resolution of 0.5°C
500 mL beakers
25 mL glass beakers
Glass stirring rod
Distilled water
Sodium hexametaphosphate (dispersing agent)
Wash bottles
Rubber stoppers to fit 1000 mL hydrometer cylinders
Stopwatch
10 mL pipette with vacuum squeeze bulb
Pipette stand

Procedure

1. Place approximately 15 gm of representative soil passing the No. 40 sieve into a small glass beaker and place the beaker in a convection oven at 110° ± 5°C.
2. After 24 h, remove the beaker and specimen from the oven and place the beaker in a desiccator to cool. After cooling, weigh out 10.00 gm of oven-dry soil and pour the soil into a 250 mL glass beaker.

3. Mix the soil to a thick slurry using 125 mL of a 40 g/liter solution of sodium hexametaphosphate and distilled water as necessary. Stir until the soil is thoroughly wetted.

4. Cover the beaker with a glass plate or plastic wrap to prevent any foreign particles from entering the solution and allow the solution to soak for a minimum of 16 hours.

5. Prepare a reference liquid consisting of distilled water and dispersing agent in the same proportions as in the sedimentation test.

6. Transfer the slurry to a malt mixer and add 125 mL of distilled water and mix for 1 minute. Take care to not lose any portion of the sample during transfer. (Note: If using the Air-Jet Dispersion apparatus, transfer the soil solution to a 1000 mL hydrometer cylinder and fill with distilled water to 250 mL. Adjust the air pressure to 172 kPa (25 psi) and disperse for 5 min.)

7. Transfer the slurry to a 1 liter hydrometer cylinder and add distilled water as necessary to bring the total volume to 1 liter. (Note: If using the Air-Jet Dispersion apparatus, carefully remove the Air-Jet and use a wash bottle of distilled water to wash all soil particles from the Air-Jet into the cylinder.)

8. Mix the solution for 1 minute by placing a rubber stopper or a hand over the open end of the hydrometer cylinder and then turning the cylinder upside down and back. Set the cylinder down and simultaneously start a timer ($t = 0$). At about the same time, make sure that the reference solution is also mixed for 1 minute using the same procedure.

9. Move the pipette over the top of the suspension and lower it until the tip just touches the surface of the suspension.

10. About 15 seconds before a sample is due, carefully lower the pipette to a depth of 10 cm below the surface of the suspension. Collect samples at different times using Table 4.8.

11. Transfer the collected samples into a 25 mL glass beaker of known mass. Place the beaker and samples into a convection oven and dry for 24 h at 110°C.

12. After oven drying, remove the beakers and determine the mass to the nearest 0.0001 gm.

13. Collect a 10 ml sample of dispersing agent solution and place it in a 25 mL glass beaker of known mass. Place the sample in the oven and dry for 24 h. After 24 h, remove the beaker and determine the dry mass of the dispersing agent.

14. After collecting the last sample, carefully shake the suspension and then pass the suspension through a No. 200 sieve to separate out the sand fraction. Use a wash bottle to wash the sand from the sieve into a small glass beaker. Place the beaker in a convection oven. After 24 h, remove the beaker and determine the oven-dry mass of the sand fraction.

Table 4.8 Sampling Times for Pipette Analysis for 10 cm Depth

	Particle size			
Specific Gravity	0.075 mm (sec)	0.020 mm (min:sec)	0.010 mm (min:sec)	0.002 mm (h:min)
2.50	20	4:35	18:20	7:38
2.55	19	4:26	17:45	7:23
2.60	18	4:18	17:12	7:09
2.65	18	4:10	16:40	6:56
2.70	17	4:03	16:11	6:44
2.75	17	3:56	15:43	6:33
2.80	16	3:49	15:17	6:22
2.85	16	3:43	14:52	6:11

Sampling times for other particle sizes may be obtained from:

$$t = \left(18 \times 10^8 \eta H\right) \big/ \left[(G-1)\left(981 \rho_w D^2\right)\right] \qquad (4.7)$$

where:

t = time (sec)
η = viscosity of water (gm/cmsec)
H = sampling depth (cm)
ρ_w = unit density of water (gm/cm^3)
D = particle diameter (μm)
G = Specific Gravity of the soil

Equation 4.7 can also be used to solve for D:

$$D = \left[(30\eta H)/(G-1)(981\rho_w H)\right]^{0.5} \qquad (4.8)$$

Calculations

The percentage, P, of particles with diameter smaller than D is obtained from:

$$P = \left(V_T/V\right)\left(M_S - M_B - M_D\right)/\left(M_0\right) \times 100\% \qquad (4.9)$$

where:

V_T = total volume of the suspension (mL)
V = volume of the pipette (ml) (generally 10 mL)
M_0 = initial mass of oven dry test specimen (gm)
M_S = mass of beaker and dried soil for each specimen collected (gm)
M_B = mass of beaker (gm)
M_D = mass of dispersing agent per 10 ml of solution (gm)

Example

The results in Table 4.9 were obtained from a sample of loess from central Iowa with an initial oven-dry mass of 12.7922 gm.
  For this soil:

  Dry mass of soil retained on the No. 200 sieve = 0.0895 gm.
  Sand = (0.0895 gm/12.7922 gm) × 100% = 0.8%
  Clay = 26.2%
  Fine silt = 74.1% − 26.2% = 47.9%
  Coarse silt = 100% − 0.8% − 26.2% − 47.9% = 25.1%

Discussion

The Pipette method and the Hydrometer method give comparable results, just in a different way. Many laboratories are set up to perform one of the tests on multiple samples at the same time, as shown in Figure 4.4. Errors

Table 4.9 Pipette Results for a Sample of Loess

Beaker no.	Particle size (mm)	Mass of beaker (gm)	Mass of beaker + dry soil (gm)	Mass of dry soil (gm)	Percent
16	.020	31.0713	31.3277	0.2541	74.1
17	.002	28.0429	28.1405	0.0976	26.2
18	>0.075	28.2541	28.3389	0.0848	0.7

Figure 4.4 Performing Hydrometer Tests on several samples by staggering reading times.

Figure 4.5 Photo of mixer and mixing cup (left) and Air-Jet Dispersion apparatus (right).

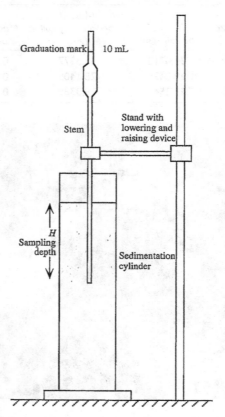

Figure 4.6 Schematic of pipette and stand.

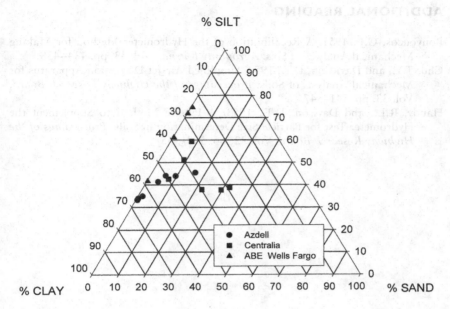

Figure 4.7 Triangular grain-size diagram.

in these tests usually occur because of errors in determining the mass of the different components of the test. The use of the Air-Jet Dispersion apparatus eliminates the need to transfer the soil from a mixing cup to the hydrometer or pipette cylinder since Dispersion is performed in the 1000 mL glass cylinder (Figures 4.5 and 4.6).

For many fine-grained soils that only have three size components, for example sand, silt and clay, it may be convenient to display the results of the hydrometer or pipette analysis in terms of a triangular (ternary) grain-size diagram, as shown in Figure 4.7. Each apex of the triangle represents 100% of the size indicated at the base of that side of the triangle. Each side of the triangle represents 0% and lines leading to the apex and parallel to the base represent different percentages – 10%, 20%, 30%, etc.

REFERENCES

Kilmer, V.J., and Alexander, L.T., 1949. Methods of Making Mechanical Analysis of Soils. *Soil Science*, Vol. 68, pp. 16–24.

Lutenegger, A.J., 2019. *Soils and Geotechnology in Construction*. CRC Taylor & Francis Publishers.

Walter, N.F., Hallberg, G.R., and Fenton, T.E., 1978. Particle-Size Analysis by the Iowa State University Soil Survey Laboratory. Standard Procedures for Evaluation of Quaternary Materials in Iowa, Iowa Geological Survey Bureau Technical Information Series No. 8.

ADDITIONAL READING

Bouyoucos, G.J., 1951. A Recalibration of the Hydrometer Method for Making Mechanical Analysis of Sols. *Agronomy Journal*, Vol. 43, pp. 434–438.

Chu, T.Y., and Davidson, D.T., 1953. Simplified Air-Jet Dispersion Apparatus for Mechanical Analysis of Soils. *Proceedings of the Highway Research Board*, Vol. 33, pp. 541–547.

Handy, R.L., and Davison, D.T., 1953. A Pipette Method to Supplement the Hydrometer Test for Particle Size Determination in Soils. *Proceedings of the Highway Research Board*, Vol. 32, pp. 548–555.

Chapter 5

Organic Content

BACKGROUND

Organic matter in soils is usually detrimental to engineering behavior and performance. In the field, organic soils can often be distinguished from inorganic soils by their characteristic odor and their dark gray to black color, but a visual assessment of organic materials may be very misleading in terms of engineering analysis. Even a small amount of organic matter in a soil can adversely affect the engineering properties and behavior. Therefore, it is important to determine the Organic Content whenever possible if the soils at a site are suspected of having organics. Some marine deposits may have high Organic Content. One of the worst organic deposits for building on is peat, a deposit consisting of decayed organic matter such as leaves, grass, twigs and other plant matter.

In general, organic soils are not considered suitable bearing material for most structures, even lightly loaded structures. Highly organic soils, such as peats, are usually first identified in the field by very dark brown to black colors, organic odor, and the presence of decaying plant matter such as grass, leaves, twigs, etc. They are often fibrous and are often lightweight. When exposed to air, they often oxidize very quickly, turning from black to brown in just a few minutes.

Most highly organic soils occur in low topographic positions or depressions that hold water. The water table is usually very high, which accounts for the presence of the organic materials. Generally, in an oxidizing environment above the water table, the organic materials have decayed and will be lost. Often topographic closed depressions in areas near lakes, rivers and other bodies of water or in glacial pothole areas are locations where some organic deposits might be encountered. ASTM D653 describes peat as "a naturally occurring highly organic substance derived primarily from plant materials". It is typically composed of vegetable tissue in various stages of decomposition, usually with an organic odor, dark brown to black color, a spongy consistency and a fibrous texture. Most peat deposits have low pH, which means they may be corrosive to untreated steel.

DOI: 10.1201/9781003263289-6

Peats and organic fine-grained soils are sometimes associated with specific geologic areas or deposits, such as glaciated regions of the upper Midwest of the U.S., Canada, Russia and Northern Europe and alluvial, lacustrine and some marine deposits around the world. A special type of organic soil, Muskeg, is a peat material often found in Arctic or northern climates which can reach depths of 100 ft, depending on the underlying topography. Organic soils and peat also can occur along coastal plain areas, where they can be up to 200 ft thick (e.g., Serva & Brunamonte 2007). Soft mud formed by the decay of peat is called *gyttja* and is found in many Scandinavian countries. The Organic Content of organic soils can vary widely, from around 2% to nearly 100% for some peat deposits. Natural Water Content can typically be over 200% and often will be as high as 400%. In some cases, Water Contents as high as 1500% have been reported (Landva & Pheeney 1980; Lefebvre et al. 1984).

In the laboratory, specific guidelines can be used to classify soils as "organic". The amount of organic material is obtained by burning off the organic matter using a muffle furnace, which can produce temperatures up to 700°C. At high temperature, the organic matter burns and leaves an ash residue. Peat is distinguished from other organic soil materials by its lower ash content (less than 25% ash by dry mass). A standard procedure for determining organic matter is given in ASTM D2974-14 *Standard Test Methods for Moisture, Ash and Organic Matter of Peat and Other Organic Soils*.

There is no standardized system for classification of organic soils based on the measured amount of organics. Within the U.S., some state highway departments have developed suggested terminology for describing organic soils. For example, the Ohio DOT uses the percentage of organic matter as determined by the loss on ignition test to describe organic soils according to Table 5.1.

According to ASTM D2487-06, the terms *organic silt* or *organic clay* can be used if the Liquid Limit of a sample obtained after over drying at 110°C overnight (LL_{OD}) is less than 75% of the same sample determined from the initial natural moisture condition (LL_{NAT}), i.e., without oven drying (Liquid Limit ratio = LL_{OD}/LL_{NAT}).

Soils with relatively high Organic Content can retain water, resulting in high Water Content, high primary and secondary compressibility

Table 5.1 Description of Organic Soils by Ohio DOT

% Organic Content	Description
2–4	Slightly organic
4–10	Moderately organic
>10	Highly organic

and potentially high corrosion potential. Organic soils may or may not be relatively weak depending on the nature of the organic material. Highly organic fibrous peats can exhibit high strengths despite having a very high compressibility, but even so they are usually avoided for construction since they are generally highly compressible.

There have been several methods suggested for determining the Organic Content of soils. Two methods are described in this chapter: 1) A loss-on-ignition (LOI) test using a muffle furnace to burn the organic matter and produce ash; and 2) digestion of organic matter using hydrogen peroxide. Both methods are considered direct test methods and essentially determine the loss of mass as the organic matter is removed. Other methods involving dry or wet combustion have also been suggested (e.g., Schmidt 1970), but these are less common for engineering applications. An automated dry combustion method is also sometimes used, especially by agronomists, using a LECO carbon analyzer (e.g., Tabatabai & Bremner 1970; Wang & Anderson 1998). This method uses a high temperature combustion in a stream of purified oxygen, and the amount of carbon dioxide (CO_2) given off is collected. Typically, a specimen with a mass of less than 1 gm is used with the LECO analyzer.

ORGANIC CONTENT BY LOSS-ON-IGNITION (LOI)

Introduction

A common laboratory test for determining the percentage of organic material in a specimen involves heating a sample to temperatures of $440° \pm 40°C$ in a muffle furnace and holding this temperature until no further change in mass occurs (ASTM D2974). At this temperature, the sample turns to ash. The specimen used for the test is a previously oven-dried specimen taken immediately after oven drying from a moisture content evaluation. Figure 5.1 shows a muffle furnace used for determining Organic Content. This method is sometimes called *Loss-On-Ignition* (LOI).

Standard

ASTM D2974 *Standard Test Methods for Moisture, Ash, and Organic Matter of Peat and Other Organic Soils*

Equipment

Electronic scale with a minimum capacity of 500 gm and a resolution of 0.01 gm
Convection oven capable of maintaining a temperature of $110° \pm 5°C$
Muffle furnace capable of maintaining a temperature of $440° \pm 40°C$
Ceramic bowls
Tongs

Figure 5.1 Muffle furnace used to determine Organic Content.

Procedure

1. Select a representative specimen of soil.
2. Select a clean dry ceramic bowl and determine the mass (M_{BOWL}) to the nearest 0.01 gm.
3. Place about 10 gm of the soil specimen in the bowl and determine the total mass ($M_{WET} + M_{BOWL}$) to the nearest 0.01 gm.
4. Place the uncovered bowl in a convection oven set at $110° \pm 5°C$ for 24 h.
5. After 24 h, remove the bowl and determine the dry mass ($M_{DRY} + M_{BOWL}$) to the nearest 0.01 gm.
6. Place the uncovered bowl in the muffle furnace set at $440°C$.
7. After 24 h, remove the bowl and determine the mass of the bowl and ash ($M_{ASH} + M_{BOWL}$) to the nearest 0.01 gm.
8. Dispose of the soil in an appropriate waste container.

Calculations

From the convection oven, the dry mass of soil, $M_{DRY} = (M_{WET} - M_{BOWL})$
The mass of water, $M_{WATER} = (M_{WET} + M_{BOWL}) - (M_{DRY} + M_{BOWL})$
From the muffle furnace, the mass of ash $= (M_{ASH} + M_{BOWL}) - M_{BOWL}$

Calculate the Water Content of the soil as:

$$W = (M_{WATER} / M_{DRY}) \times 100\%$$

Calculate the ash content of the soil as:

$$\text{Ash Content} = AC = (M_{ASH}/M_{DRY}) \times 100\%$$

Calculate the Organic Content of the soil as:

$$OC = 100\% - AC$$

Example

A sample of organic clay was collected at a site. The following values were determined in the laboratory:

$$M_{WET} + M_{BOWL} = 156.35 \text{ gm}$$
$$M_{BOWL} = 97.66 \text{ gm}$$
$$M_{DRY} + M_{BOWL} = 138.49 \text{ gm}$$
$$M_{ASH} + M_{BOWL} = 136.88 \text{ gm}$$

Determine the Water Content and Organic Content.

$$M_{DRY} = (M_{DRY} + M_{BOWL}) - M_{BOWL} = 138.49 - 97.66 = 40.83 \text{ gm}$$
$$M_{WATER} = (M_{WET} + M_{BOWL}) - (M_{DRY} + M_{BOWL}) = 156.35 \text{gm} - 138.49 \text{ gm} = 17.86 \text{ gm}$$
$$W = (M_{WATER}/M_{DRY}) \times 100\% = (17.86 \text{ gm}/40.83 \text{ gm}) \times 100\% = 43.7\%$$
$$\text{Ash Content (AC)} = (M_{ASH}/M_{DRY}) \times 100\%$$
$$M_{ASH} = 136.88 - 97.66 = 39.22 \text{ gm}$$
$$AC = (39.22 \text{ gm}/40.83 \text{ gm}) \times 100\% = 96.1\%$$
$$OC = 100\% - AC = 100\% = 96.1\% = 3.9\%$$

Discussion

Rankin (1970) had presented a method for determining organic matter in soils based on a simple titration method. The method actually determines the organic carbon and assumes that:

$$\text{Organic Matter} = 1.72 \times \text{Organic Carbon}$$

Several investigations have suggested that results obtained by the loss-on-ignition method can be influenced by both the ignition temperature and time of heating (e.g., Howard 1966; Spain et al. 1982; David 1988; Schulte et al. 1991). The combustion of organic matter usually only takes 5–6 h, but most laboratories keep the specimen in the muffle furnace at least overnight or 24 h just for convenience. Recently, Joy et al. (2021) found that LOI can be adopted for all soils regardless of the amount of organic matter content present.

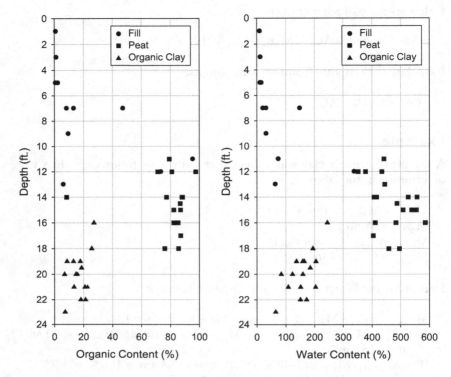

Figure 5.2 Organic Content and Water Content in fill, peat and organic clay at a site in Pittsfield, Ma.

Figure 5.2 shows the Organic Content and Water Content determined using the LOI procedure at a site in Pittsfield, Ma. The soil profile consisted of several feet of random fill overlying a peat layer of about 2.4 m (8 ft) in thickness to a depth of about 5.5 m (18 ft) overlying soft organic clay. The peat layer shows an Organic Content of about 80% while the underlying organic clay has an Organic Content of between about 10% and 20%. Note that the Water Content throughout the profile generally shows that soils with higher Organic Content also have higher Water Content. Also note that the Water Content in the peat layer is nearly 600% since the dry mass of the organic material is so low.

ORGANIC CONTENT BY HYDROGEN PEROXIDE

Introduction

Determination of Organic Content by the hydrogen peroxide method is based on the digestion of organic matter by an alkaline solution. The specimen is treated with hydrogen peroxide (H_2O_2) to oxidize organic matter resulting in a loss of mass. The organic matter is then determined by comparing the mass of material loss to the original mass of the specimen.

The hydrogen peroxide method was suggested for determining Organic Content in the early 1900s, and the test procedure has changed very little since then. The test is based on oxidation (digestion) of organic matter, which is a basic chemical reaction.

Standard

There currently is no ASTM standard test method for determining Organic Content using the hydrogen peroxide method.

Equipment

Electronic scale with a minimum capacity of 500 gm and a resolution of 0.01 gm

Convection oven capable of maintaining a temperature of $110° \pm 5°C$

Ceramic bowl or 250 mL glass beaker

Tongs

Glass stirring rod

Procedure

1. Select a representative specimen of soil.
2. Prepare a solution of 30% or 50% hydrogen peroxide (H_2O_2) with distilled water.
3. Select a clean dry ceramic bowl or a 250 mL glass beaker and determine the mass (M_{BOWL}) to the nearest 0.01 gm.
4. Place the soil specimen in the bowl and determine the total mass ($M_{WET} + M_{BOWL}$) to the nearest 0.01 gm.
5. Place the uncovered bowl in a convection oven set at 110°C for 24 h.
6. After oven drying, remove the bowl and determine the dry mass ($M_{DRY} + M_{BOWL}$) to the nearest 0.01 gm.
7. Add 10 mL of H_2O_2 solution to the bowl or beaker and mix gently with a glass stirring rod to digest the organic matter. Wash any soil adhered to the glass rod into the bowl with H_2O_2.
8. Continue adding H_2O_2 in 10 mL increments until bubbling ceases.
9. After digestion is complete, place the bowl and soil in the convection oven and dry at 110°C for 24 h.
10. After oven drying, remove the bowl and determine the mass of the bowl and ash ($M_{FINAL} + M_{BOWL}$) to the nearest 0.01 gm.
11. Dispose of the soil in an appropriate waste container.

Calculations

From the convection oven, the initial dry mass of soil,
$$M_{DRY} = (M_{WET} + M_{BOWL}) - M_{BOWL}$$
After digestion and oven drying, $= M_{FINAL} = (M_{FINAL} + M_{BOWL}) - M_{BOWL}$

Calculate the Organic Content of the soil as:

$$OC = (M_{DRY} - M_{FINAL}) / M_{DRY} \times 100\%$$

Example

A sample of organic clay soil from Cambridgeshire, UK, was tested for Organic Content.

Initial dry mass of bowl No. B10 $M_{BOWL} = 126.65$ gm
Initial wet mass of soil plus bowl $= M_{WET} + M_{BOWL} = 175.30$ gm
After oven drying, $M_{DRY} + M_{BOWL} = 148.92$
Initial dry mass of soil $= 148.92$ gm $- 126.65$ gm $= 22.27$ gm
After digestion and oven drying, $M_{FINAL} + M_{BOWL} = 144.33$ gm
$M_{FINAL} = 144.33$ gm $- 126.65$ gm $= 17.68$ gm
Mass of organic matter digested $= 22.27$ gm $- 17.68$ gm $= 4.59$ gm
Organic Content $= (4.59$ gm$/17.68$ gm$) \times 100\% = 26.0\%$

Discussion

The hydrogen peroxide method is simple to perform, like the loss-on-ignition test, but does not require a muffle furnace. It only uses basic laboratory equipment. However, the oxidation of organics may not be complete. This method is sometimes used to pre-treat soils prior to performing the grain-size analysis to remove organic matter.

According to ASTM D2487-06, the terms *organic silt* or *organic clay* can be used if the Liquid Limit of a sample obtained after oven drying at 110°C overnight is less than 75% of the Liquid Limit of the same sample determined from the initial natural moisture condition, i.e., without oven drying (Liquid Limit Ratio $= LL_{OD}/LLNAT$). Figure 5.3 shows results of tests performed on soils in Indiana both before and after oven drying. There is a clear trend of decreasing Liquid Limit Ratio with increasing Organic Content. However, for soils containing more than about 40% organics, the determination of the Atterberg Limits is difficult and the Water Content within the soil is not necessarily distributed evenly, as the organics tend to contain higher amounts of water that the actual soil mineral particles. Test methods for determining Liquid Limit are discussed in Chapter 12.

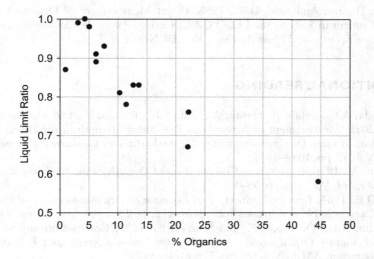

Figure 5.3 Relationship between Liquid Limit Ratio and Organic Content (data from Huang et al. 2009).

REFERENCES

David, M.B., 1988. Use of Loss-on-Ignition to Assess Soil Organic Carbon in Forest Soils. *Communication Soil Science Plant Analysis*, Vol. 19, pp. 1593–1599.

Howard, P.J., 1966. The Carbon-Organic Matter Factor in Various Soil Types. *Oikos*, Vol. 15, pp. 229–236.

Huang, P.T., Patel, M., Santagata, M., and Bobet, A., 2009. Classification of Organic Soils. Report No. FHWA/IN/JTRP-2008/2, Indiana Department of Transportation and Federal Highway Administration.

Joy, A., Abraham, B.M., and Sridharan, A., 2021. A Critical Re-Examination of the Factors Influencing Determination of Organic Matter in Soils. *Geotechnical and Geological Engineering*, Vol. 39, pp. 4287–4293.

Rankin, W.L., 1970. Suggested Method of Test for Organic Matter Content of Soils by Redox Titration. *ASTM STP*, Vol. 479, pp. 286–287.

Schmidt, N.O., 1970. Suggested Method of Test for Organic Carbon Content of Soil by Wet Combustion, *ASTM STP*, Vol. 479, pp. 271–278.

Spain, A.V., Probert, M.E., Isbell, R.F., and John, R.D., 1982. Loss-on-Ignition and the Carbon Content of Australian Soils. *Australian Journal of Soil Research*, Vol. 20, pp. 147–152.

Schulte, E.E., Kaufman, C., and Peter, J.B., 1991. The Influence of Sample Size and Heating Time on Soil Weight Loss on Ignition. *Communication Soil Science Plant Analysis*, Vol. 22, pp. 159–168.

Tabatabai, M.A., and Bremner, J.M., 1970. Use of the LECO Automatic 70-Second Carbon Analyzer for Total Carbon Analysis of Soils. *Soil Science Society of America Proceedings*, Vol. 34, pp. 608–610.

Wang, D., and Anderson, D.W., 1998. Direct Measurement of Organic Carbon Content in Soils by the LECO CR-12 Carbon Analyzer. *Communications in Soil Science and Plant Analysis*, Vol. 19, Nos. 1 & 2, pp. 15–21.

ADDITIONAL READING

Amunda, A.G., Sahdi, F., Hasan, A., Taib, S.N., Boylan, N., and Mohamad, A., 2019. Measurement of Amorphous Peat Shear Strength in the Direct Shear Box at High Displacement Rates. *Geotechnical and Geological Engineering*, Vol. 37, pp. 1059–1072.

Arman, A., 1970. Engineering Classification of Organic Soils. *Highway Research Record*, Vol. 310, pp. 75–89.

Bell, D.F., 1964. Loss on Ignition as an Estimate of Organic Matter and Organic Carbon in Noncalcareous Soils. *Journal of Soil Science*, Vol. 15, pp. 84–92.

Franklin, A.G., Orozco, L.F., and Semrau, R., 1973. Compaction and Strength of Slightly Organic Soils. *Journal of the Soil Mechanics and Foundations Division, ASCE*, Vol. 99, No. 7, pp. 541–557.

Gould, R.A., Bedell, P.R., and Muckle, J.G., 2002. Construction Over Organic Soils in an Urban Environment: Four Case Histories. *Canadian Geotechnical Journal*, Vol. 39, No. 2, pp. 345–356.

Gunaratne, M., Stinnette, P., Mullins, A.G., Kuo, C.L., and Echelberger, W., 1998. Compressibility Relations for Peat and Organic Soils. *Journal of Testing and Evaluation, ASTM*, Vol. 26, No. 1, pp. 1–9.

Kogure, K., and Ohira, Y., 1977. Statistical Forecasting of Compressibility of Peaty Ground. *Canadian Geotechnical Journal*, Vol. 14, No. 4, pp. 562–570.

Landva, A., and Pheeney, P.E., 1980. Peat Fabric and Structure. *Canadian Geotechnical Journal*, Vol. 17, No. 3, pp. 416–435.

Lefebre, G., Langlois, P., Lupien, C., and Lavallee, J., 1984. Laboratory Testing and in Situ Behavior of Peat as Embankment Foundation. *Canadian Geotechnical Journal*, Vol. 21, No. 2, pp. 322–337.

Li, W., O'Kelly, B.C., Yang, M., Fang, K., Li, X., and Li, H., 2020a. Briefing: Specific Gravity of Solids Relationship With Ignition Loss for Peaty Soils. *Geotechnical Research*, Vol. 7, No. 3, pp. 134–145.

Malkawi, A.I., Alawneh, A.S., and Abu-Safaqah, O.T., 1999. Effects of Organic Matter on the Physical and the Physicochemical Properties of an Illitic Soil. *Applied Clay Science*, Vol. 14, pp. 257–278.

O'Kelly, B.C., 2015. Atterberg Limits are not Appropriate for Peat Soils. *Geotechnical Research, ICE*, Vol. 2, No. 3, pp. 123–134.

Robinson, W.O., 1927. The Determination of Organic Matter in Soils by Means of Hydrogen Peroxide. *Journal of Agricultural Research*, Vol. 34, No. 4, pp. 339–356.

Serva, L., and Brunamonte, F., 2007. Subsidence in the Pontina Plain, Ital. *Bulletin of Engineering Geology and the Environment*, Vol. 66, pp. 125–134.

Skempton, A.W., and Petley, D.J., 1970. Ignition Loss and Other Properties of Peats and Clays from Avonmouth, King's Lynn and Cranberry Moss. *Geotechnique*, Vol. 20, No. 4, pp. 343–356.

Walker, A., 1947. A Critical Examination of a Rapid Method for Determining Organic Carbon in Soils: Effect of Variation in Digestion Conditions and of Inorganic Soil Constituents. *Soil Science*, Vol. 37, pp. 251–263.

Zentar, R., Abriak, N.E., and Dubois, V., 2009. Fall Cone Test to Characterize Shear Strength of Organic Sediments. *Journal of Geotechnical and Geoenvironmental Engineering*, Vol. 135, No. 1, pp. 153–157.

Chapter 6

Carbonate Content

BACKGROUND

Carbonate minerals (calcium carbonate [$CaCO_3$] and calcium-magnesium carbonate [$MgCaCO_3$]) may be present in many geologic deposits and can act as cementing agents to hold individual particles or clusters of particles together, influencing soil behavior. The presence of carbonates in varying degrees can influence both the classification characteristics as well as the behavioral characteristics of soils and may help explain unusual behavior in a geologic deposit or differences in behavior within the same geologic deposit. The amount and type of carbonates present in soils depends largely on carbonates present in the original deposit or parent material and also depends on any addition or removal by precipitation or leaching that may have occurred after deposition.

Carbonate composition may be useful to geotechnical engineers in several ways, including: 1) evaluating effect on geotechnical behavior; 2) identifying individual geologic strata; 3) identifying leached/severely modified zones; 4) identifying cemented strata; and 5) identifying a source of bonding in soils. The amount and type of carbonates present may exert substantial influence on the engineering properties and behavior. Numerous studies have documented the presence of carbonates in both onshore and offshore soil deposits (e.g., Quigley 1980; Burghignoli et al. 1991; Boone & Lutenegger 1997) and detailed studies have demonstrated that carbonates can influence a variety of geotechnical behavioral characteristics of fine-grained soils, ranging from Atterberg Limits (e.g., Samuels 1975; Datta & Sultania 1982; Hawkins et al. 1988; Cotecchia & Chandler 1995) to compressibility and stress history (e.g., Boone & Lutenegger 1997; Bozzano et al. 1999) and residual shear strength (e.g., Hawkins & McDonald 1992; Fukue et al. 1999).

The actual amount and type of carbonate present in each soil depends on the geologic origin and the degree of alteration it has undergone, e.g., leaching, precipitation, etc. Some soils have 0% carbonates while some marls can contain up to 70% carbonates by mass. There are two popular methods to determine Carbonate Content. One method is performed in a closed

DOI: 10.1201/9781003263289-7

pressure vessel and measures the pressure of CO_2 gas produced when carbonates react with weak hydrochloric acid. The other method measures the volume of CO_2 gas produced from this reaction. The basic reaction between the acid and carbonates is:

$$CaCO_3 + HCl \text{ gives } CaCl_2 + H_2O + CO_2 \qquad (6.1)$$

Quantitative determination of the amounts of carbonates is typically performed using a laboratory gasometric technique in which dry soil is subjected to digestion by dilute hydrochloric acid in a closed vessel (ASTM D4373) or by determining the amount of CO_2 gas produced (Dreimanis 1962). In the author's opinion, either of the tests is simple enough that the determination of carbonates should be considered as a part of routine soil characterization.

CARBONATE CONTENT BY RAPID CARBONATE ANALYZER

Introduction

The measurement of total Carbonate Content in soil is standardized by ASTM D4373 which uses a pressure vessel (referred to as the *Rapid Carbonate Analyzer*) to measure the pressure of CO_2 gas produced when dilute hydrochloric acid reacts with carbonates in the soil. The procedure is largely based on the work by Williams (1948) and Muller and Gastner (1971) as described by Demars et al. (1983). A schematic of the pressure vessel is shown in Figure 6.1.

The device consists of:

1. A pressure vessel with a threaded top;
2. A 20 mL plastic container with a bail handle;
3. A threaded lid or cap for the pressure vessel;
4. An O-ring casket between the vessel and lid;
5. A bourdon pressure gauge;
6. A pressure relief valve.

Since the gas pressure is generated by the reaction of HCl and carbonates within the soil low, a Lucite, acrylic or comparable inert plastic material may be used to fabricate the Carbonate Analyzer. Using a clear plastic also allows the operator to see that the acid and soil have mixed properly. Commercial models of the Carbonate Analyzer are also available, but a local machine shop can easily fabricate the device. Figure 6.2 shows a photo of the Carbonate Analyzer fabricated and used by the author.

Standard

ASTM D4373 *Standard Test Method for Calcium Carbonate of Soils*

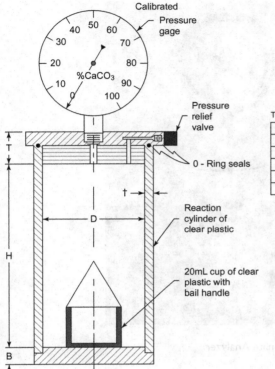

Parameter	Maximum		Minimum	
B	0.50	(13.)	0.25	(6.4)
D	3.0	(76.)	1.5	(38.0)
H	5.5	(140.)	4.0	(100.)
T	1.25	(32.)	0.75	(19.0)
†	0.375 (10.)		0.25	(6.4)

Typical range of dimensions, in. (mm)

Figure 6.1 Schematic of Rapid Carbonate Analyzer. (after ASTM D4373)

Equipment

Electronic scale with a minimum capacity of 100 gm and a resolution of
     0.01 gm
Carbonate Analyzer
1N hydrochloric acid
Reagent grade $CaCO_3$
Stopwatch
Convection oven capable of maintaining $110° \pm 5°C$

Calibration of the Carbonate Analyzer

Each Carbonate Analyzer must be calibrated with known amounts of
$CaCO_3$ to account for any differences in dimensions resulting from the fab-
rication. The pressure gauge on the Carbonate Analyzer should be selected
to give the approximate range of maximum pressure that might be encountered in soils. The pressure gauge should have a minimum resolution of 0.2

Figure 6.2 Photo of Rapid Carbonate Analyzer.

kPa (1.5 psi). Calibrate the Carbonate Analyzer with samples of $CaCO_3$ by using 10 replicate samples of 0.1 gm mass, beginning at 0.1 gm and continuing to 1.0 gm.

Prepare a 1N solution of hydrochloric acid by placing 80 mL of concentrated HCl in a 1 L volumetric flask. Dilute by adding distilled water to the 1L mark.

1. Place a calibration specimen in the bottom of the Carbonate Analyzer.
2. Place 20 ml of 1 N hydrochloric acid in the acid container and set the container in the bottom of the Carbonate Analyzer, being careful not to spill any acid.
3. Check that the pressure gauge on the lid of the Carbonate Analyzer reads zero. If not, record the pressure gauge reading as R_0.
4. Open the relief valve on the lid of the Carbonate Analyzer and tighten the lid onto the top.
5. Close the relief valve and slowly tilt the Carbonate Analyzer to initiate the reaction between the acid and the soil. Start a stopwatch at the same time. Mix the contents using a slight swirling action.
6. Record the pressure gauge reading at 30 sec as R.

7. Open the relief valve to release the gas pressure. Remove the lid and properly dispose of the contents. Clean and dry the equipment for another test.
8. Prepare a calibration chart as shown in Figure 6.3.

Procedure

1. Select a representative soil sample. Prepare the sample by passing the material through a No. 40 sieve. Oven dry the specimen for 24 h in a convection oven. Obtain a 1.00 gm specimen of oven dry soil for testing, M_{DRY}. (Note: If low Carbonate Content is suspected, 2 gm of soil may be used in the test.)
2. Perform the test by using the procedure described previously.
3. If the pressure gauge continues to increase after 30 sec, allow the reaction to continue for 35 min. At the end of 35 min, record the pressure reading as R_{35}.
4. After obtaining the pressure reading, release the valve on the top cap. Remove the soil and dispose of the material in an appropriate container.

Calculations

Use the maximum pressure gauge reading adjusted for the initial reading $(R - R_0)$ and determine the mass of total carbonates using the calibration chart as M_C.

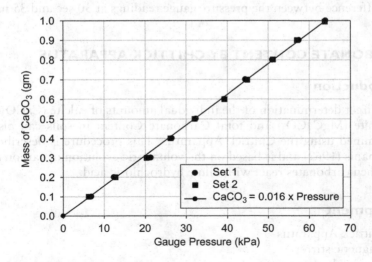

Figure 6.3 Rapid Carbonate Analyzer calibration chart.

Calculate total carbonates as:

$$\%\text{Carbonates} = \left(M_C / M_{DRY}\right) \times 100\% \tag{6.2}$$

Report the % carbonates to the nearest 0.1%

Example

A 2 gm oven-dried sample of clay from Louisiana was tested for carbonates using the Carbonate Analyzer. $R_0 = 0.2$ kPa. A pressure reading of 14.2 kPa was obtained after 30 sec. The calibration of the Carbonate Analyzer is given in Figure 6.3.

$R = 14.2$ kPa

The corrected pressure is $P = 14.2$ kPa $- 0.2$ kPa $= 14.0$ kPa. From the correlation shown in Figure 6.3, $CaCO_3 = 0.016$ $P = 0.016 \times 14.0$ kPa $= 0.224$ gm

$$\%\text{Carbonates} = \left(0.224\,\text{gm}/2\,\text{gm}\right) \times 100\% = 11.2\%$$

Discussion

Total carbonates are expressed as a percentage of the dry mass of the soil. In soils with low carbonates, the mass of the sample can be increased to 2 gm to improve the reliability of the measurement. Soil passing the No. 40 sieve is recommended as carbonates are important in fine-grained soils. Tests on sands may be performed on material passing the No. 10 sieve if appropriate. ASTM D4373 notes that if dolomite is present, the reaction will continue for 30–40 min or more. The dolomite content may be estimated by taking the difference between the pressure gauge readings at 30 sec and 35 min.

CARBONATE CONTENT BY CHITTICK APPARATUS

Introduction

The direct determination of the individual amounts of calcite ($CaCO_3$) and dolomite ($MgCaCO_3$) and total Carbonate Content in soils can also be determined using the Chittick Apparatus. This procedure is described by Dreimanis (1962) and is based on the volumetric evolution of carbon dioxide when carbonates react with dilute hydrochloric acid.

Equipment

Chittick Apparatus
Magnetic stirrer
Stopwatch
Flasks

Electronic scale with a minimum capacity of 100 gm and a resolution of
  0.01 gm
Dilute hydrochloric acid
Barometer
Thermometer with a range of 0–50°C and a resolution of 0.1°C
Convection oven capable of maintaining 110°±5°C
Sodium chloride
Methyl orange powder
Sodium bicarbonate

Setup

A schematic of the Chittick Apparatus is shown in Figure 6.4. An oven-dry
soil sample of known mass is placed in the flask (B) and rests on a magnetic
stirrer (A). The flask is connected to an adjustable graduated tube and fluid
reservoir (D), which contains a colored fluid to make the readings of the gas
volume easier. Dilute hydrochloric acid is introduced into the sample flask
using a graduated burette (C). As carbon dioxide is generated from the reaction of the acid with soil carbonates, the fluid in the reservoir is displaced.

Two solutions are required for the test: 1) colored reservoir fluid; and 2)
dilute hydrochloric acid.

Reservoir fluid

The reservoir fluid should be a sharp pink color. If the fluid is clear, yellow
or faint pink, drain the reservoir and replace with fresh solution. The reservoir fluid is prepared as follows:

1. In a 1L or 2L Erlenmeyer Flask, dissolve 100.0 g of sodium chloride
   (NaCl) in 350 mL of distilled water.
2. Add 1.0 g of sodium bicarbonate.
3. Add 2 mL of methyl orange solution of 15.0 mg of methyl orange
   powder.
4. Add 1:5 dilute sulfuric acid (1 part concentrated sulfuric acid to 5
   parts distilled water) until the solution turns a deep pink (usually
   about 10 mL).
5. Stir overnight.
6. Add distilled water to fill up to 1 L.
7. Stir for 1 hour.

Dilute hydrochloric acid

Prepare a 6N solution of HCl by adding 109.4 gm of concentrated HCl to
1 liter of distilled water.

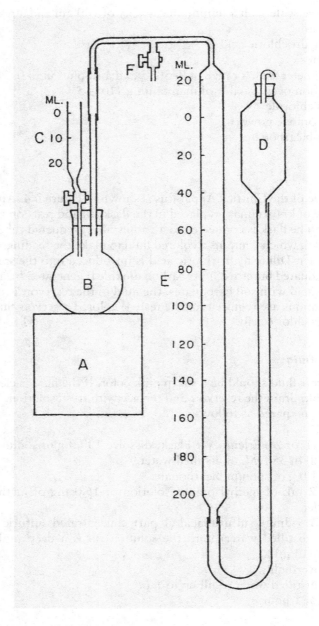

Figure 6.4 Schematic of Chittick apparatus.

Procedure

The test is performed as follows:

1. Select a representative sample of soil passing the No. 40 sieve. Place about 5–10 gm of the soil in an aluminum tare in the oven overnight at 110° ±5°C.
2. Remove the sample from the oven; weigh out a sample of 1.70 g for the test.
3. Switch on the magnetic stirrer at least 10 min before testing to ensure that there are no temperature effects.
4. Place the sample in a clean, dry 250 mL flask.
5. Add a plastic-coated stirring magnet to the flask.
6. Apply a thin film of vacuum grease to the inside lip of the flask, fit the rubber stopper to the top of the flask and place the flask on the stirrer.
7. Open the vent at the top of the glass "T" and raise the reservoir until the level in the graduated tube and reservoir are even at the zero line.
8. Close the vent. Lower the reservoir approximately 3 mL below the fluid level in the graduated tube. This will produce a slight negative pressure in the system.
9. At the same time, start the stopwatch and open the valve on the acid burette. Close the valve after 20 ml of acid has drained into the flask.
10. Read the volume of CO_2 evolved after 1 min by quickly leveling the fluid in the graduated tube and the reservoir. Subtract 20 mL from this reading to account for the volume of gas displaced from the 20 ml of HCl. Record this value as R_1. Also record the temperature and barometric pressure.
11. Lower the reservoir again to maintain a slight negative pressure as before.
12. Read the volume of CO_2 evolved after 20 min by again leveling the fluid. Subtract 20 ml from this reading and record this value as R_{20}. Again, record the temperature and barometric pressure.
13. Turn off the magnetic stirrer, open the valve at the top of the "T", and bring the reservoir level back to zero. Remove the flask, clean and dry it, and prepare another sample for testing.

Calculations

Reduce the data from the test as follows:

Look up the correction factors for temperature and barometric pressure for each of the readings (see Appendix II).

Multiply R_1 and R_{20} by their respective correction factors to obtain CR_1 and CR_{20}.

Calculate the percentages (based on 1.70 g of oven dried soil) of calcite and dolomite as:

$F = CR_1 - (0.04)(CR_{20} - CR_1)$
$E = CR_{20} - CR_1$
% calcite = 0.232 (F)
% dolomite = 0.223 (E) + 0.3
Total carbonates % = % calcite + % dolomite

Report the respective percentages of calcite, dolomite and total carbonates to the nearest 0.1%.

Example

The following results were obtained from a sample of clay from Fargo, North Dakota, which was tested for carbonates using the Chittick Apparatus.

$R_1 = 21.0$
$R_{20} = 78.0$
$T = 26.0°C$
$P = 769$ mm

From Appendix II: Correction factor, CF, for 26°C and 769 mm = 1.06179

$CR_1 = R_1 \times CF = 21 \times 1.06179 = 22.3$
$CR_{20} = R_{20} \times CF = 78.0 \times 1.06179 = 82.8$
$F = 22.3 - (0.04)(82.8 - 22.3) = 19.9$
$E = 82.8 - 22.3 = 60.5$
% calcite = 0.232(19.9) = 4.6%
% dolomite = 0.223 (60.5) + .3 = 13.8%
Total carbonates = 4.6% + 13.8% = 18.4%

Discussion

In the field, an indication of the presence of carbonates can be obtained qualitatively by applying a few drops of dilute hydrochloric acid to a soil sample and observing the speed and vigor of the reaction. However, the reaction also depends on the Water Content of the soil and on the presence of other minerals. Such observations are useful and should be recorded on the field boring log and should be included in a soil description to note that the soil is either "calcareous" or "leached". Table 6.1 gives some typical ranges of Carbonate Content for several different geologic deposits. Total carbonates includes both calcite and dolomite. It may be useful to report all three quantities in a report. Figure 6.5 shows the variation in calcite and dolomite at two sites measured with the Chittick Apparatus.

Table 6.1 Typical Ranges of Total Carbonate Content for Different Soils

Location	Geology	Range of measured total carbonates (%)
Boston, Ma.	Boston Blue Clay	2.8–16.1
Massena, NY	Leda Clay	10.4–21.0
Amherst, Ma.	Connecticut Valley Varved Clay	2.0–6.6
Albany, NY	Glacial Lake Clay	14.4–27.1
Evanston, Ill.	Chicago Clay	22.2–31.8
Clinton County, Ia.	Loess	2.9–21.8
Ames, Ia.	Glacial Till	5.5–16.5
London, UK	London Clay	2.2–8.8
Centralia, Mo.	Glacial Clay and Till	2.1–8.8
Houston, Texas	Beaumont and Montgomery Clay	2.1–25.2
Salt Lake City, Utah	Lacustrine Clay	13.6–34.4
Treasure Island, Ca.	San Francisco Bay Mud	2.8–4.3

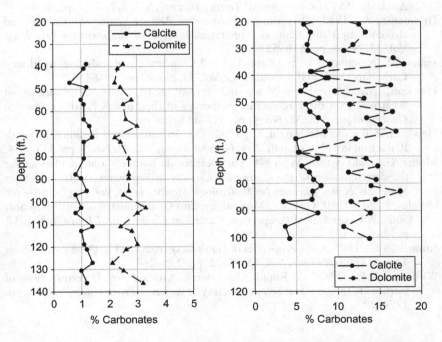

Figure 6.5 Variation in calcite and dolomite at two sites: Boston Blue Clay, Boston, Ma. (left); Albany, NY (right).

REFERENCES

Boone, S.J., and Lutenegger, A.J, 1997. Carbonates and Cementation of Glacially Derived Cohesive Soils in New York State and Southern Ontario. *Canadian Geotechnical Journal*, Vol. 34, No. 4, pp. 534–550.

Bozzano, F., Marcoccia, S., and Barbieri, M., 1999. The Role of Calcium Carbonate in the Compressibility of Pliocene Lacustrine Deposits. *Quarterly Journal of Engineering Geology*, Vol. 32, pp. 271–289.

Burghignoli, A., Cavalera, L., Chieppa, V., Jamiolkowski, M., Mancuso, C., Marchetti, S., Pane, V., Paoliani, P., Silvestri, F., Vinale, F., and Vittori, E., 1991. Geotechnical Characterization of Fucino Clay. *Proceedings of the 10th European Conference on Soil Mechanics and Foundation Engineering*, Vol. 1, pp. 27–40.

Cotecchia, F., and Chandler, R.J., 1995. Geotechnical Properties of the Pleistocene Clays of the Pappadai Valley, Taranto, Italy. *Quarterly Journal of Engineering Geology*, Vol. 28, pp. 5–22.

Datta, M., and Sultania, S.N.K., 1982. Nature and Engineering Properties of Fine Grained Calcareous Soils Found at Two Sites on the Western Continental Shelf of India. *Proceedings of the 2nd Canadian Conference on Marine Geotechnical Engineering*.

Demars, K.R., Chaney, R.C., and Richter, J.A., 1983. The Rapid Carbonate Analyzer. *ASTM Geotechnical Testing Journal*, Vol. 6, No. 1, pp. 30–34.

Dreimanis, A., 1962. Quantitative Gasometric Determination of Calcite and Dolomite by Using Chittick Apparatus. *Journal of Sedimentary Petrology*, Vol. 32, No. 3, pp. 520–529.

Fukue, M., Nakamura, T., and Kato, Y., 1999. Cementation of Soils Due to Calcium Carbonate. *Soils and Foundations*, Vol. 39, No. 6, pp. 55–64.

Hawkins, A.B., Lawrence, M.S., and Privett, K.D., 1988. Implications of Weathering on the Engineering Properties of the Fuller's Earth Formation. *Geotechnique*, Vol. 38, No. 4, pp. 517–532.

Hawkins, A.B., and McDonald, C., 1992. Decalcification and Residual Strength Reduction in Fuller's Earth Clay. *Geotechnique*, Vol. 42, No. 3, pp. 453–464.

Muller, G., and Gastner, M., 1971. The "Karbonate-Bombe", a Simple Device for the determination of the Carbonate Content in Sediments, Soils and Other Materials. *Neues Jahrbuch fur Mineralogic Monatschefte*, Vol. 10, pp. 466–469.

Quigley, R. M., 1980. Geology, Mineralogy, and Geochemistry of Canadian Soft Soils: A Geotechnical Perspective. *Canadian Geotechnical Journal*, Vol. 17, pp. 261–285.

Samuels, S.G., 1975. Some Properties of Gault Clay from the Ely-Ouse Essex Water Tunnel. *Geotechnique*, Vol. 25, No. 2, pp. 239–264.

Williams, D.E., 1948. A Rapid Manometric Method for Determination of Carbonate in Soils. *Soil Science Society of America Proceedings*, Vol. 13, pp. 127–129.

ADDITIONAL READING

Bascomb, C.L., 1961. A Calcimeter for Routine Use on Soil Samples. *Chemistry and Industry*, Vol. 45, pp. 1826–1827.

Beyhan, S., and Emir, E., 2010. Effect of Calcium Carbonate on Engineering Properties of Marl. *Proceedings of the ISRM International Symposium – EUROROCK 2010.*

Collins, S.H., 1906. Scheibler's Apparatus for the Determination of Carbonic Acid in Carbonates: An Improved Construction and Use for Accurate Analysis. *Journal of the Society of Chemical Industry*, Vol. 25, June 15, pp. 518–522.

Dunn, D.A., 1980. Revised Techniques for Quantitative Calcium Carbonate Analysis Using the "Karbonate-Bombe", and Comparisons to Other Quantitative Carbonate Analysis Methods. *Journal of Sedimentary Petrology*, Vol. 50, pp. 631–636.

Evangelou, V.P., Wittig, L.D., and Tanji, K.K., 1984. An Automated Manometric Method for Quantitative Determination of Calcite and Dolomite. *Soil Science Society of America Journal*, Vol. 48, pp. 1236–1239.

Fukue, M., and Nakamura, T., 1996. Effects of Carbonate on Cementation of Marine Soils. *Marine Georesources and Geotechnology*, Vol. 14, No. 1, pp. 37–45.

Fukue, M., and Nakamura, T., 1999. Cementation of Soils Due to Calcium Carbonate. *Soils and Foundations*, Vol. 39, No. 6, pp. 55–64.

Imai, G., Komatsu, Y., and Fukue, M., 2006. Consolidation Yield Stress of Osaka-Bay Pleistocene Clay with Reference to Calcium Carbonate Contents. *Journal of ASTM International*, Vol. 3, No. 7, pp. 89–97.

Jones, G.A., and Kaiteris, P., 1983. A Vacuum-Gasometric Technique for Rapid Precise Analysis of Calcium Carbonate in Sediments and Soils. *Journal of Sedimentary Petrology*, Vol. 53, pp. 655–658.

Lamas, F., Irigary, C., and Chacon, J., 2002. Geotechnical Characterization of Carbonate Marls for the Construction of Impermeable Dam Cores. *Engineering Geology*, Vol. 66, pp. 283–294.

Lamas, F., Irigary, C., Oteo, C., and Chacon, J., 2005. Selection of the Most Appropriate Method to Determine the Carbonate Content for Engineering Purposes with Particular Regard to Marls. *Engineering Geology*, Vol. 81, pp. 32–41.

Martin, A.E., and Reeve, R., 1955. A Rapid Manometric Method for Determining Soil Carbonate. *Soil Science*, Vol. 79, pp. 187–197.

Nakamura, T., Kukue, M., and Naoe, K., 1993. Effects of Calcium Carbonate on Geotechnical Properties of Sediments in Tokyo Bay. *Proceedings of the 3rd International Offshore and Polar Engineering Conference*, Paper ISOPE-I-93-094.

O'Neill, M. W., 2000. National Geotechnical Experimentation Site – University of Houston. *National Geotechnical Experimentation Sites, ASCE*, pp. 72–101.

Presley, B.J., 1975. A Simple Method for Determining Calcium Carbonate in Sediment Samples. *Journal of Sedimentary Petrology*, Vol. 45, No. 3, pp. 745–746.

Samuels, S.G., 1975. Some Properties of Gault Clay from the Ely-Ouse Essex Water Tunnel. *Geotechnique*, Vol. 25, No. 2, pp. 239–264.

Schink, J.C., Stockwell, J.H., and Ellis, R.A., 1979. An Improved Device for Gasometric Determination of Carbonate in Sediment. *Journal of Sedimentary Research*, Vol. 49, No. 2, pp. 651–653.

Skinner, S.I.M., and Halstead R.L., 1958. Note on Rapid Method for Determination of Carbonates in Soils. *Canadian Journal of Soil Science*, Vol. 38, pp. 187–188.

Smolik, L.C., 1936. New Apparatus for Volumetric Determination of Carbonates and Water Content in Soils. *Proceedings of the 1st International Conference on Soil Mechanics and Foundation Engineering*, Vol. 3, pp. 47–48.

Wolfe, J.A., and Bartlett, V.C., 1958. Gasometric Determination of Calcite and Dolomite. *Geological Society of America Bulletin*, Vol. 69, pp. 1664–1665.

Woodward, L., 1961. A Manometer Method for the Rapid Determination of Lime in Soils. *Soil Science Society of America Proceedings*, pp. 248–250.

Chapter 7

pH of Soils

INTRODUCTION

Soil pH is a simple measurement that requires little equipment and can give some insight into soil composition through acidity and alkalinity. The measure of acidity or alkalinity of soils in a solution of distilled/deionized water is given as pH, which is a measure of the hydrogen ion concentration. pH is defined as the logarithm of the reciprocal of the hydrogen ion concentration, i.e., pH = log(1/H$^+$); where H$^+$ = moles of H$^+$ per L. For each whole number decrease in pH, there is an increase in the number of hydrogen ions present by a factor of 10. The pH of soils may provide a preliminary indication of the soil's corrosivity as it might affect steel, such as used in driven piles, buried pipes, anchors, etc. Solutions with a pH less than 7 are acidic, and solutions with pH greater than 7 are basic. Pure water has a pH of 7 or neutral:

$$pH = \log(1/0.0000001) = \log(1000000) = 7$$

These tests are simple to perform, and results are obtained in about an hour. pH is measured on a 1:1 soil–water mixture and should be measured at room temperature, approximately 20°C. The complete test procedure measures pH in both a soil–water solution and a soil–0.01M calcium chloride solution. According to ASTM D4972, both liquids are necessary to fully define the soil's pH; however, many laboratories only perform pH using distilled water. A standard test method for determining the pH of water is given in ASTM D1293.

STANDARD

ASTM D4972-13 *Standard Test Method for pH of Soils*; ASTM D2976 *Standard Test Method for pH of Peat Materials*

DOI: 10.1201/9781003263289-8

EQUIPMENT

100 mL glass beaker
Electronic scale with a minimum capacity of 250 gm and a resolution of
  0.01 gm
Glass stirring rod
pH meter with glass electrode (temperature compensated)
100 mL graduated cylinder
Standard reference buffer solutions of pH = 4.0, 7.0 and 10.0
1L volumetric flask

PROCEDURE

1. Calibrate the pH meter using the manufacturer's operating instructions using standard pH buffer solutions of 4.0, 7.0 and 10.0.
2. Select a representative sample of air-dried soil to be tested that has
   been sieved through a No. 10 (2 mm) sieve.
3. Weigh out 25 gm of soil and place in a clean 100 mL glass beaker.
4. Add 20 mL of distilled-deionized water to the beaker and mix thoroughly with a clean glass rod.
5. Let the soil–water mixture stand for 1 h.
6. Follow the manufacturer's instructions for operating the pH meter
   and immerse the glass electrode into the solution and read the pH on
   the pH meter (Figure 7.1).
7. Report the pH to the nearest 0.1
8. Using a clean 100 mL glass beaker, determine the pH of the distilled-
   deionized water used in the test.
9. Report the pH of the distilled-deionized water to the nearest 0.1.
10. If pH measurements are to be made using a 0.01M calcium chloride
    solution, first prepare the solution. Prepare a 1.00M stock calcium
    chloride solution by dissolving 147 gm of $CaCl_2*2H_2O$ in water in a
    1L volumetric flask. Dilute to 1L with water. Dilute 20 mL of the 1.0
    M $CaCl_2$ solution to 2L with water.
11. Repeat the procedure using 0.01M calcium chloride solution instead
    of distilled-deionized water to prepare a soil–calcium chloride
    solution.
12. Report the pH to the nearest 0.1.

CALCULATIONS

There are no calculations for pH. Measurements are taken directly from
the pH meter.

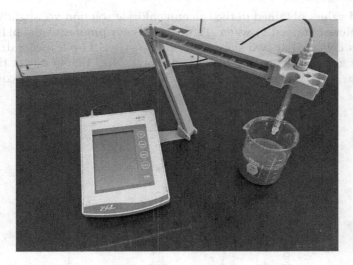

Figure 7.1 Typical laboratory electronic direct read pH meter.

DISCUSSION

It is important to frequently check the calibration of the pH meter with standard buffer solutions to make sure that the meter is functioning properly. Also, proper care of the probe is important. The probe should be routinely cleaned, and when not in use it should be stored in the buffer solution so that it doesn't dry out.

Soil pH is useful in determining the solubility of soil minerals and is often used to give an approximate indication of the corrosivity of the soil. Figure 7.2 shows the range of soil pH and the ranges for high corrosion potential. Soil pH depends on several factors, including the type of soil, the grain size, the mineralogy and the composition of the pore water.

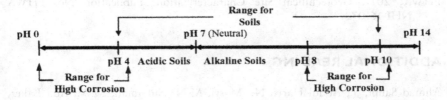

Figure 7.2 Ranges of soil pH for high corrosivity potential (from FHWA 2017).

Typical soils have pH values ranging from about 5.5 to 7.5, although some soils may clearly be outside this range. The pH of sea water is in the range of 7.5 to 8.4. pH measurements in the calcium chloride solution displaces some of the exchangeable aluminum present in most soils.

The pH values obtained in the calcium chloride solution are slightly lower than values obtained in water. Figure 7.3 shows profiles of the pH of soil samples collected at two sites where shallow steel H-piles were driven to support ground-mount PV solar panel arrays. Note that at one of the sites (AFSolar), pH values are acidic, which may suggest long-term corrosion issues.

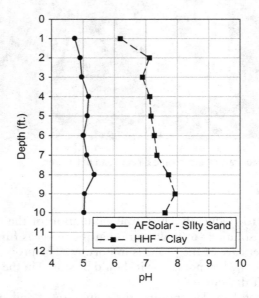

Figure 7.3 pH measurements obtained at two sites.

REFERENCE

FHWA, 2017. Geotechnical Site Characterization, Publication No. FHWA NHI-16-072.

ADDITIONAL READING

Ahmad Saupi, S., Abdul Haris, N., Masri, M.N., Sulaiman, M.A., Abu Baker, M.B., Mohamad Amani, M., Mohamed, M., and Nik Yusuf, N.A., 2016. Effects of Soil Physical Properties to the Corrosion of Underground Pipelines. *Materials Science Forum*, Vol. 840, pp. 309–334.

Decker, J.B., Rollins, K., and Ellsworth, J.C., 2008. Corrosion Rate Evaluation and Prediction for Piles Based on Long-Term Performance. *Journal of Geotechnical and Geoenvironmental Engineering, ASCE*, Vol. 134, No. 3, pp. 341–351.

Doyle, G., Seica, M., and Grabinsky, M., 2013. The Role of Soil in the External Corrosion of Cast Iron Water Mains in Toronto, Canada. *Canadian Geotechnical Journal*, Vol. 40, No. 2, pp. 225–236.

Ferreira, C.A., Ponciano, J., Vaitsman, D.S., and Perez, D.V., 2007. Evaluation of the Corrosivity of the Soil Through its Chemical Composition. *Science of the Total Environment*, Vol. 388, No. 1, pp. 250–255.

Ikechukwu, A.S., Ugochukwu, N.H., Ejimofor, R.A., and Obioma, E., 2014. Correlation Between Soil Properties and External Corrosion Growth Rate of Carbon Steel. *The International Journal of Engineering and Science*, Vol. 3, No. 10, pp. 38–47.

Moore, T., and Hallmark, C., 1987. Soil Properties Influencing Corrosion of Steel in Texas Soils. *Soil Science Society of America Journal*, Vol. 51, No. 5, pp. 1250–1256.

Wong, I.H., and Law, K.H., 1999. Corrosion of Steel H Piles in Decomposed Granite. *Journal of Geotechnical and Geoenvironmental Engineering, ASCE*, Vol. 125, No. 6, pp. 529–532.

Chapter 8

Cation Exchange Capacity

INTRODUCTION

There is often a negative charge at the surface of clay crystals because of isomorphous substitution and imperfections in the crystal lattice. This occurs because in many clay minerals, an atom of a lower positive valence replaces an atom of higher valence, resulting in a deficient positive charge. This leads to unsatisfied valence charges at the edges of the crystal. The degree of this deficiency means that the clay wants to attract other cations to become electrically neutral. Different clays have different charge deficiencies and therefore have different tendencies to attract cations. This deficiency is referred to as *Cation Exchange Capacity* (CEC) and essentially describes the degree to which a particular soil wants to attract more cations. This means that, effectively, CEC is a measure of the negative surface charge on the mineral surface. CEC is generally expressed in units of milliequivalents per 100 grams of dry soil. Exchangeable cations present in most soils include calcium (Ca), magnesium (Mg), sodium (Na) and potassium (K) cations in decreasing order of abundance.

There are several suggested methods for determining CEC used by different agencies, e.g., the U.S. Department of Agriculture as described by Rhoades (1982). The exchangeable cation analysis is performed by displacing exchangeable cations (calcium, magnesium, sodium and potassium) with ammonium ions with successive washing of the soil with a neutral 1N solution of ammonium acetate and then estimating the displaced cations by atomic absorption phot-photometry or flame photometry. There are several slight variations in the test procedures, but most of the standardized test procedures for determining Cation Exchange Capacity require detailed laboratory distillation, titration and leaching procedures. These methods provide a direct chemical measurement of the CEC. Many laboratories are not equipped to routinely perform these tests, but there are several commercial laboratories that will perform the analysis at reasonably low cost. The method described in this chapter measures CEC by first displacing the index ions in the soil with another salt solution and then measuring the amount of the displaced index ions.

DOI: 10.1201/9781003263289-9

STANDARD

ASTM D7503 *Standard Test Method for Measuring the Exchange-Complex and Cation Exchange Capacity of Inorganic Fine-Grained Soils*

EQUIPMENT

Convection oven capable of maintaining $110° \pm 5°C$
No. 10 sieve
No. 40 sieve
Desiccator with silica gel desiccant
Electronic scale with a minimum capacity of 20 gm and a resolution of 0.001 gm
Small ceramic weighing dish
Shaker
Spectrophotometer
500 mL filtering flask with connection for flexible tubing to low pressure vacuum pump
Vacuum pump
Flexible tubing
Buchner funnel (55 mm or 90 mm diameter)
Wash bottles
100 mL graduated cylinder
Ashless filter paper (2.5 μm) to fit over Buchner funnel
250 mL volumetric flasks
Distilled water
Isopropanol

REAGENTS

1M ammonium acetate
Prepare an ammonium acetate solution by dissolving 77.08 gm of 99.9% pure NH_4OAc in distilled water and fill to 1000 mL in a volumetric flask. Adjust the pH of the solution to 7 with ammonium hydroxide or acetic acid. (Note: Approximately 1L of NH_4OAc is needed per six samples.)
1 M potassium chloride
Prepare potassium chloride solution by dissolving 74.6 gm of 99.9% pure KCl in distilled water and fill to 1000 mL in a volumetric flask. (Note: Approximately 1L on KCl is needed per six samples.)
Ammonium sulfate
Prepare ammonium sulfate solution by drying 238 mg of ACS Certified $(NH_4)_2SO_4$ for 4 hrs. at 40°C. Make a 200 gm/L stock solution by

dissolving the dried compound in 100 mL of distilled water and then fill to volume in a 205 mL Erlenmeyer flask. Prepare calibration standards by diluting the solution into concentrations of 10, 20, 40, 50 and 80 mL.

PROCEDURE

1. Select a representative sample of soil passing the No. 10 sieve (or No. 40 sieve) and air dry approximately 40 gm of soil. (Note: 12 gm of soil is needed for the test.)
2. Determine the Water Content of the air-dried soil using about 10 gm of soil using the convection oven method described in Chapter 1.
3. Using the calculated Water Content, determine the total mass needed to give 10 gm of dry soil mass.
4. Rinse an acid washed 500 mL filtering flask with isopropanol.
5. Place 10 gm of soil and 40 mL of 1M NH_4OAc into a 100 mL covered container that will fit into an end-over-end shaker.
6. Shake the covered container for 5 min at 5 rpm. Agitate the container to rinse any particles from the sides of the container and then let the container sit for 24 h.
7. After 24 h, shake the container for 15 min at 30 rpm.
8. Rinse the 500 mL filtering flask and Buchner funnel with 1 M NH_4OAc.
9. Place the Buchner funnel over the 500 mL filtering flask and line the Buchner funnel with 2.5 μm ashless filter paper.
10. Transfer the contents of the shaken container to the Buchner funnel.
11. Rinse the container and cap into the acid funnel using 1M NH_4OAc.
12. Apply low suction (< 10 kPa [1.4 psi]) to the filtering flask.
13. Wash the soil in the Buchner funnel with four 30 mL portions of 1 M NH_4OAc. Add each 30 mL portion slowly and allow each 30 mL portion to drain before adding the next portion. (Note: Do not allow the soil to dry between additions of NH_4OAc solution. After the last washing, turn off the vacuum.)
14. Place the Buchner funnel with 1 M NH_4OAc washed sample into the 500 mL filtering flask.
15. Apply low suction (< 10 kPa [1.4 psi]) to the filtering flask. (Note: Do not allow the soil to dry when suction is applied.)
16. Wash the soil with three 40 mL portions of isopropanol. Allow each 40 mL portion to drain before adding the next portion.
17. Turn off the suction to the filtering flask when free liquid is no longer visible.
18. Separate the Buchner funnel from the filtering flask and discard the isopropanol collected in the 500 mL filtering flask and then rinse the flask with distilled water three times.

19. Place the Buchner funnel containing the isopropanol washed soil onto the rinsed filtering flask.
20. Apply low suction to the filtering flask again and wash the soil with four 50 mL portions of 1M KCl solution. Allow each portion of KCl solution to drain before adding the next portion. (Note: Do not allow the soil to dry between additions of KCl solution.)
21. Rinse a 250 mL volumetric flask with 1M KCl.
22. Transfer the extract into the 250 mL volumetric flask. Rinse the filtering flask with distilled water and transfer the contents into the volumetric flask.
23. Fill the volumetric flask to volume with distilled water.
24. Analyze the KCl extract for nitrogen concentration (mg/L) using a spectrophotometer.

CALCULATIONS

Calculate the Cation Exchange Capacity (CEC) as:

$$CEC = N \left(1\,cmol^+/140\,mg\right)\left(0.25\,L/M_S\left(1000\,g/kg\right)\right) \tag{8.1}$$

where:

CEC = Cation Exchange Capacity (cmol$^+$/kg = meq/100 gm)
N = concentration of nitrogen (from Step 24) (mg/L)
M_{DRY} = oven-dry mass of soil

DISCUSSION

The test procedure presented in this chapter may be challenging for many laboratories to perform. The materials require careful handling as with any chemicals and take up considerable counter space if performing tests on multiple samples at the same time. There are alternatives to this procedure. For example, Miller et al. (1975) described a simple procedure using an ammonia electrode for determining CEC. Typically values of CEC for different fine-grained soils are given in Table 8.1.

Note that CEC does not apply to coarse-grained soils, which effectively do not have a charge deficiency. Cation Exchange Capacity tends to be lowest in kaolinites and highest in montmorillonites. Within a given clay mineral suite, CEC and SSA are related, as discussed in Chapter 9.

A parameter that can be used to give a preliminary indication of a soil's clay mineralogy is CEC Activity, obtained as A_{CEC} = CEC/(% < 0.002 mm.) (Cerato & Lutenegger 2005). This is similar to Skempton's Activity described in Chapter 12, but is based on the measured CEC. Table 8.2 gives some values of A_{CEC} for different clay minerals.

Table 8.1 Typical Measured Values of Cation Exchange
Capacity of Several Fine-Grained Soils

Soil	Cation Exchange Capacity (meq/100 gm)
Peerless Clay (Kaolinite)	1
Theile Kaolin	2
Texas Ca Montmorillonite	84
Wyoming Bentonite	76
Boston Blue Clay	15
Leda Clay	18
Houston Clay	29
Salt Lake City Clay	20
Omaha Loess	17
London Clay	43
Osaka Japan Clay	21
Gault Clay	27

Table 8.2 Cation Exchange Activity for Different
Clay Minerals

Clay	A_{CEC}
Kaolinite (well ordered)	0.06
Peerless Clay (K)	0.01
Kaolinite (poorly ordered)	0.05
Thiele Kaolin (K)	0.03
Illite Green shale	0.37
Old Hickory Clay (I)	0.57
Palygorskite	0.29
NaCa Montmorillonite	1.28
Hectorie	0.82
Ca Montmorillonite	1.51
Na Montmorillonite	1.26
Southern Bentonite	1.35
Ca Montmorillonite	3.18

REFERENCES

Cerato, A.B., and Lutenegger, A.J., 2005. Activity, Relative Activity and Specific Surface Area of Fine-Grained Soils. *Proceedings of the 16th International Conference on Soil Mechanics and Foundation Engineering*, Vol. 2, pp. 325–328.

Miller, G.A., Rieken, F.F., and Walter, N. F., 1975. Use of an Ammonia Electrode for Determination of Cation Exchange Capacity in Soil Studies. *Soil Science Society of America Proceedings*, Vol. 39, pp. 372–373.

Rhoades, J.D., 1982. Cation Exchange Capacity. In *Methods of Soil Analysis. Part 2: Chemical and Microbiological Properties Agronomy 9*, pp. 149–157. Soil Science Society of America.

ADDITIONAL READING

Arthur, E., 2017. Rapid Estimation of Cation Exchange Capacity from Soil Water Content. *European Journal of Soil Science*, Vol. 68, pp. 365–373.

Chapman, H.D., 1965. Cation Exchange Capacity. In *Methods of Soil Analysis – Chemical and Microbiological Properties. Agronomy*, Vol. 9, pp. 891–901.

Davidson, D.T., and Sheeler, J.B., 1952. Cation Exchange Capacity of Loess and its Relation to Engineering Properties. *ASTM STP*, Vol. 142, pp. 1–19.

Khorshidi, M., and Lu, N., 2017. Determination of Cation Exchange Capacity from Soil Water Retention Curve. *Journal of Engineering Mechanics, ASCE*, Vol. 143, No. 6, p. 8.

Kolbuszewski, J., Birch, N., and Shojobi. J.O., 1965. Keuper Marl Research. *Proceedings of the 5th International Conference on Soil Mechanics and Foundation Engineering*, Vol. 1, pp. 59–63.

Ross, D.S., and Ketterings, Q., 2011. Recommended Methods for Determining Cation Exchange Capacity. In *Recommended Soil Testing Procedures for the Northeastern United States*. 3rd Edition. Northeastern Publication No. 493, University of Delaware, pp. 75–85.

Yukelen, Y., and Kaya, A., 2006. Prediction of Cation Exchange Capacity from Soil Index Properties. *Clay Minerals*, Vol. 41, No. 4, pp. 827–837.

Chapter 9

Specific Surface Area

INTRODUCTION

It is well known that surface phenomena are important factors influencing the behavior of fine-grained soils. Clay mineralogy, Cation Exchange Capacity and surface area are all important and are interrelated. Surface area can exhibit a significant influence on many physical and chemical properties of fine-grained soils. The term *Specific Surface Area* (SSA) refers to the area per unit mass of soil and is usually expressed in units of m^2/gm. In addition to Cation Exchange Capacity, SSA may be the dominant factor in controlling the behavior of many fine-grained soils. Clay particles contribute the greatest amount of surface area of any of the mineral constituents of soil but may also differ a great deal in Specific Surface Area.

There is strong evidence in the literature that indicates that Specific Surface Area may be the single most important contributing factor that controls the engineering behavior of fine-grained soils, especially the interaction of soil with water. For example, a number of other studies have shown that there is a strong relationship between Specific Surface Area and Atterberg Limits (e.g., Farrar & Coleman, 1967; Kuzukami et al. 1971; Ohtsubo et al. 1983; Locat et al. 1984; Smith et al. 1985; Morin & Dawe 1986; Churchman & Burke 1991). The swelling potential of clays has also been related to surface area as shown by several studies (e.g., Dos Santos & Castro 1965; Ross 1978; Dasog et al. 1988; Lutenegger 2022).

Several studies have also shown that surface area is one of the most important parameters influencing frost heave in soils (e.g., Anderson & Tice 1972; Reike et al. 1983; Nixon 1991). The segregation potential (SP), which describes the velocity of water arriving at an advancing frost front, has been shown to be directly related to Specific Surface Area. Surface area may also play a significant role in controlling the behavior of dispersive clays through surface charge properties (e.g., Heinzen & Arulanandan 1977; Harmse & Gerber 1988; Sridharan et al. 1992; Bell & Maud 1994). Surface area may also be significant in evaluating the effectiveness of soil stabilization, for example when treated with lime (Moore & Jones 1971).

DOI: 10.1201/9781003263289-10

Specific Surface Area varies greatly between soils because of differences in mineralogy, organic composition and particle-size distribution. For example, swelling clays such as montmorillonites have Specific Surface Areas up to 810 m^2/gm. Non-expanding soils such as kaolinites typically have Specific Surface Areas ranging from 10 to 40 m^2/gm. Consequently, the type of clay mineral present in soil is of major importance in determining the effect of Specific Surface Area on soil properties. The test procedure described in this chapter uses the ethylene glycol monoethyl ether (EGME) method based on the work of Carter et al. (1965) as modified by Cerato and Lutenegger (2002) to determine the total surface area of fine-grained soils.

STANDARD

At the present time there is no ASTM standard test procedure for determining surface area. A method for determining surface area based on adsorption of ethylene glycol was presented by ASTM Subcommittee 6 (ASTM 1970), but it appears that this procedure was never formally approved by ASTM. The method was essentially an adaptation of the method presented by Mortland and Kemper (1965).

PREPARATION

1. Prepare desiccant by placing approximately 110 gm of 40-mesh laboratory grade anhydrous calcium chloride in a clean ceramic bowl, and place the bowl in a standard convection oven set at 110°C for 1 h.
2. After 1 h, remove the calcium chloride from the oven and pour 100 gm into a 200 mL clean glass beaker.
3. Add 20 mL of EGME to the beaker and mix thoroughly with a glass stirring rod.
4. Spread the desiccant evenly in a shallow glass dish and place the dish in the bottom of the desiccator. The desiccant should always be stored in a sealed desiccator when not in use.

EQUIPMENT

Large desiccator (210–250 mm inside diameter)
Shallow aluminum tares (76 mm diameter × 25 mm height)
Electronic analytical balance with a minimum capacity of 100 gm and a resolution of 0.0001 gm
Convection oven capable of maintaining 110° ± 5°C
Shallow glass dish

Plexiglass tare lids
Glass stirring rod
200 mL glass beaker
Ceramic bowl
10 mL pipette
Vacuum grease
Vacuum pump
Ethylene glycol monoethyl ether (EGME)
Calcium chloride

Figure 9.1 shows the general arrangement of the test equipment.

PROCEDURE

1. Select a representative specimen of soil.
2. Select a clean dry aluminum tare and determine the mass to the nearest 0.0001 gm (M_{TARE}).
3. Place approximately 1 gm of oven-dried soil passing a No. 40 sieve in the bottom of the tare.
4. Determine the mass of the soil and tare ($M_{DRY} + M_{TARE}$) to the nearest 0.0001 gm.
5. Using a 10 mL pipette, gently place approximately 3 mL of laboratory grade EGME over the soil.

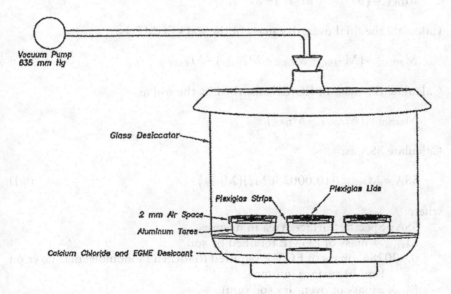

Figure 9.1 General test arrangement for determining Specific Surface Area.

6. Gently swirl the soil and EGME by hand until the mixture forms a slurry and the appearance of the slurry is uniform. (Note: Be careful not to spill any of the material and do not stir.)
7. Place the tare with soil plus EGME mixture into a vacuum desiccator and place a small Plexiglass lid over the tare, leaving a gap of 2–3 mm between the lid and the tare.
8. Apply a thin layer of vacuum grease to the bottom edge of the desiccator lid and then place the lid onto the top of the desiccator and rotate slightly to create a seal with vacuum grease.
9. Attach the top of the desiccator to a vacuum pump with a rubber hose and begin evacuating using a vacuum of at least 635 mm Hg.
10. After 8–10 h, remove the tare and determine the mass of the soil + EGME mixture (M_{DRY} + M_{EGME}). Repeat this step again after approximately 18 h and again after approximately 24 h. The mass of the mixture should not vary more than 0.001 gm. If the mass is still increasing, place the tare back in the desiccator, continue evacuating and weigh it again in approximately 4 h. Continue until the sample mass does not vary by more than 0.001 gm.

CALCULATIONS

Calculate the initial oven-dry mass of soil as:

$$M_{DRY} = (M_{TARE} + M_{DRY}) - M_{TARE}$$

Calculate the final oven-dry mass of the soil + EGME as:

$$M_{FINAL} = (M_{TARE} + M_{DRY} + M_{EGME}) - M_{TARE}$$

Calculate the mass of EGME adsorbed by the soil as:

$$M_{EGME} = (M_{FINAL} - M_{DRY})$$

Calculate SSA as:

$$SSA = M_{EGME}/[(0.000286M_S)(M_{DRY})] \tag{9.1}$$

where:
 SSA = Specific Surface Area in m^2/gm
 M_{EGME} = mass of EGME retained by soil
 0.000286 = mass of EGME required to form a monomolecular layer on 1 m^2 of surface (gm/m^2)
 M_{DRY} = mass of oven-dry soil (gm)

Report SSA to the nearest whole number.

EXAMPLE

A sample of Leda Clay from Ontario, Canada, was tested and gave the following results:

Initial mass of dry tare = 30.0485 gm
Initial mass of tare + oven-dry soil = 31.0515 gm
Initial mass of oven-dry soil = M_S = 31.0515 − 30.0485 = 1.0030 gm
Mass of tare + soil + EMGE after 18 h = 31.0634 gm
Mass of EGME after 18 h = 31.0634 − 30.0485 = 1.0149 gm
Mass of tare + soil + EMGE after 24 h = 31.0630 gm
Final mass of dry soil + EMGE after 24 h = 31.0630 − 30.0485 = 1.0145 gm
Mass of EGME = 1.0145 − 1.0030 = 0.015 gm
SSA = 0.015 gm/[(1.0030 gm)(0.000286)] = 52 m²/gm

DISCUSSION

Other test methods may be used to determine or estimate specific surface, but not all methods give the same results. For example, the nitrogen absorption method (BET) described by Brunauer et al. (1938) only gives a measure of the *external* surface area, since nitrogen does not penetrate the interlayer of clay minerals. Water vapor adsorption may also be used, and the relationship between adsorbed water and SSA is discussed in Chapter 20.

It is very important to obtain mass values with great care since the test relies on a very small mass of soil. Aluminum or glass tares are used since their mass remains constant during the evacuation and handling while weighing. A shallow tare is used since it is essential that the soil be spread thinly and evenly on the bottom of the tare to ensure that there is complete coverage of the EGME over all soil particles. At least two tests should be performed on the same soil, and SSA should be reported as the average value of the two tests.

The exact amount of EGME added to the specimen is not as important provided it is close to the recommended 3 mL. Figure 9.2 shows the time to reach equilibrium mass for the same sample with both 5 mL and 3 mL EGME added. Both tests show an equilibrium condition reached after about 30 h.

Aluminum tares with dimensions of 76 mm in diameter by 25 mm in height are used since they allow the soil to be spread adequately over the bottom and efficiently fit into the desiccators. Plexiglass lids are placed on top of each aluminum tare to prevent the soil from being pulled out of the tare during the evacuation process. Lids can be fabricated from 1 mm thick Plexiglass, and a 2 mm gap can be created between the tare and the Plexiglass lid using a thin piece of Plexiglass, as shown in Figure 9.3.

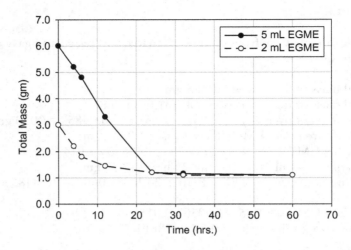

Figure 9.2 Time required to reach equilibrium mass for different amounts of added EGME.

Figure 9.3 Flat aluminum dish and plexiglass lid for SSA Measurement.

Physical and chemical properties of fine-grained soils, especially clays, may be greatly influenced by the amount of their surface areas. Fine-grained soils differ in surface area predominantly as a result of differences in texture (grain-size distribution) and types and amounts of different clay minerals.

Figure 9.4 illustrates the importance of particle size on surface area. To illustrate the influence of grain size on SSA, Table 9.1 gives the theoretical calculated surface area of cubical particles of different size but all with a total volume of 1 cm^3. Natural clay deposits can have a wide range of total

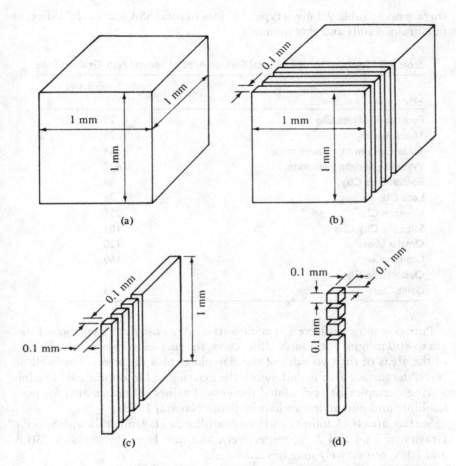

Figure 9.4 Specific Surface Area of blocks of different size (from Spangler & Handy 1982).

Table 9.1 Surface Area of Different Size Cubes in 1 cm³ of Soil

Cube size (mm)	Total volume (cm³)	Total mass (gm)	Total surface area (cm²)	Specific surface (m²/gm)
10	1	2.70	6	0.0022
1	1	2.70	60	0.022
0.1	1	2.70	600	0.22
0.01	1	2.70	6000	2.2
0.001	1	2.70	600000	22.0

surface area. Table 9.2 gives typical values of total SSA for several different fine-grained soils and clay minerals.

Table 9.2 Measured Values of Total Surface Area of Several Fine-Grained Soils

Soil	Specific Surface Area (m^2/gm)
Peerless Kaolinite Clay	23
Theile Kaolin	38
Texas Calcium Montmorillonite	534
Wyoming Sodium Bentonite	637
Boston Blue Clay	44
Leda Clay	36
Houston Clay	255
Salt Lake City Clay	101
Omaha Loess	120
London Clay	169
Osaka Japan Clay	171
Gault Clay	153

Pure clay minerals have a Specific Surface Area ranging from about 1 m^2/gm to 800 m^2/gm. For non-swelling clays, the surface area is simply the sum of the areas of the two sides of the clay plates plus the edges. For swelling clays, the surface area includes both the exterior and interior areas. The following examples give calculated theoretical values of surface area for pure kaolinite and pure montmorillonite (from Nagaraj 1993).

Surface areas of kaolinite and montmorillonite clay minerals with Specific Gravity of 2.64 and 2.76, respectively, and unit layer thickness or 750 Å and 10 Å, respectively may be claculated.

Kaolinite

Specific Gravity = mass/volume = 1 gm/V

Volume = 1/G (cm^3/gm)
$a \times b \times c = 1/G$
$a \times b = 1/Gc$

Specific Surface Area = 2 × area on either side of particles:

$2(a \times b) = 2/Gc$

Kaolinite G = 2.64 and assuming c = 750 Å

$$SSA = \left[1/10^4\right]\left[\left(2 \times 10^8\right)/\left(2.64 \times 0.75 \times 10^3\right)\right] = 10\,m^2/gm$$

Montmorillonite

Montmorillonite $G = 2.76$ and assuming well dispersed state $c = 10$ Å

$$SSA = \left[1/10^4 \right]\left[\left(2 \times 10^8 \right)/\left(2.76 \times 10\right) \right] = 725 \, m^2/gm$$

A parameter that can be used to give a preliminary indication of clay mineralogy is Surface Area Activity, obtained as $A_{SSA} = SSA/(\% < 0.002 \, mm)$. This is similar to Skempton's Activity, described in Chapter 12, and CEC Activity, previously described in Chapter 8, but is based on the measured SSA. Table 9.3 gives some values of A_{SSA} for different clay minerals (Cerato & Lutenegger 2005).

Within a given deposit of unique geology, it should be expected that CEC and SSA are related. This has generally been found to be the case, except that other factors may influence the relationship and therefore some scatter should be expected. In addition, most natural fine-grained deposits do not contain pure clay minerals, and most clays are composed of a combination of mixed mineral particles. Other soil compositional constituents, such as Carbonate Content and Organic Content, may also influence the relationship between CEC and SSA. Figure 9.5 shows the relationship between CEC and SSA for pure clay mineral mixes – sodium montmorillonite (NaM) plus kaolinite (K) and calcium montmorillonite (CaM) plus kaolinite (K) soil samples collected at three different sites – and demonstrates the complexity of fine-grained soils. Figure 9.6 shows the relationship between SSA and CEC for three natural fine-grained deposits.

Table 9.3 Surface Area Activity for Different Clay Minerals

Clay	A_{SSA}
Kaolinite (well ordered)	0.41
Peerless Clay (K)	0.30
Kaolinite (poorly ordered)	0.38
Thiele Kaolin (K)	0.49
Illite Green shale	3.29
Old Hickory Clay (I)	5.23
Palygorskite	5.07
NaCa Montmorillonite	8.04
Hectorie	7.25
Ca Montmorillonite	7.29
Na Montmorillonite	10.55
Southern Bentonite	13.64
Ca Montmorillonite	20.34

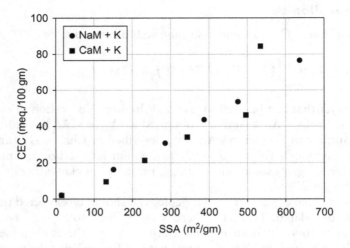

Figure 9.5 Relationship between CEC and SSA for two clay mineral mixes.

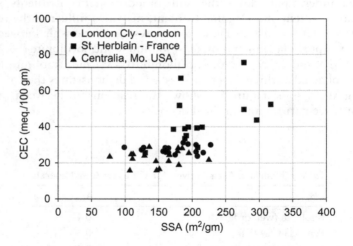

Figure 9.6 Relationship between CEC and SSA for three natural fine-grained deposits.

REFERENCES

Anderson, D.M., and Tice, A.R., 1972. Predicting Unfrozen Water Contents in Frozen Soils from Surface Area Measurements. *Highway Research Record*, Vol. 393, pp. 12–18.

ASTM, 1970. Suggested Method for Determining Specific Surface. *ASTM STP*, Vol. 479, pp. 279–285.

Bell, F.G., and Maud, R.R., 1994. Dispersive Soils: A Review from a South African Perspective. *Quarterly Journal of Engineering Geology*, Vol. 27, pp. 195–210.

Brunauer, S., Emmett, P.H., and Teller, E., 1938. Adsorption of Gases in Multi-Molecular Layers. *Journal of the American Chemical Society*, Vol. 60, pp. 309–319.

Carter, D.L., Heilman, M.D., and Gonzalez, C.I., 1965. Ethylene Glycol Monoethyl Ether for Determining Surface Area of Silicate Minerals. *Soil Science*, Vol. 100, No. 5, pp. 356–360.

Cerato, A.B., and Lutenegger, A.J., 2002. Recommended Method for Determining Surface Area of Fine-Grained Soils. *Geotechnical Testing Journal, ASTM*, Vol. 25, No. 3, pp. 315–321.

Cerato, A.B., and Lutenegger, A.J., 2005. Activity, Relative Activity and Specific Surface Area of Fine-Grained Soils. *Proceedings of the 16th International Conference on Soil Mechanics and Foundation Engineering*, Vol. 2, pp. 325–328.

Churchman, G.J., and Burke, C.M., 1991. Properties of Subsoils in Relation to Various Measures of Surface Area and Water Content. *Journal of Soil Science*, Vol. 42, pp. 463–478.

Dasog, G.S., Acton, D.F., Mermut, A.R., and DeJong, E., 1988. Shrink-Swell Potential and Cracking in Clay Soils of Saskatchewan. *Canadian Journal of Soil Science*, Vol. 68, pp. 251–260.

Dos Santos, M.P.P., and DeCastro, E., 1965. Soil Erosion in Roads. *Proceedings of the 6th International Conference on Soil Mechanics and Foundation Engineering*, Vol. 1, pp. 116–118.

Farrar, D.M., and Coleman, J.D., 1967. The Correlation of Surface Area with Other Properties of Nineteen British Clay Soils. *Journal of Soil Science*, Vol. 18, No. 1, pp. 118–124.

Harmse, H.J., and Gerber, F.A., 1988. A Proposed Procedure for the Identification of Dispersive Soils. *Proceedings of the 2nd International Conference on Case Histories in Geotechnical Engineering*, Vol. 1, pp. 411–416.

Heinzen, R.T., and Arulanandan, K., 1977. Factors Influencing Dispersive Clays and Methods of Identification. *ASTM STP*, Vol. 623, pp. 202–217.

Kuzukami, H., Ozaki, E., and Nakaya, M., 1971. Relationships Between Specific Surface and Liquid Limit. *Transactions of the Japanese Society of Irrigation, Drainage and Reclamation Engineering*, Vol. 37, pp. 61–67.

Locat, J. Lefebvre, G., and Ballivy, G., 1984. Mineralogy, Chemistry, and Physical Property Interrelationships of Some Sensitive Clays from Eastern Canada. *Canadian Geotechnical Journal*, Vol. 21, pp. 530–540.

Lutenegger, A.J., 2022.

Moore, J.C., and Jones, R.L., 1971. Effect of Soil Surface Area and Extractable Silica, Alumina, and Iron on Lime Stabilization Characteristics of Illinois Soils. *Highway Research Record*, Vol. 351, pp. 97–92.

Morin, P., and Dawe, C.R., 1986. Geotechnical Properties of Two Deep Sea Marine Soils from Labrador Sea Area. *Proceedings of the 3rd Canadian Conference on Marine Geotechnical Engineering*, Vol. 1, pp. 117–134.

Mortland, M.M., and Kemper, W.D., 1965. Specific Surface. Chapter 42. In C.A. Black (Ed.), *Methods of Soil Analysis*, pp. 523–544. American Society of Agronomy.

Nagaraj, T.S., 1993. *Principles of Testing Soils, Rocks and Concrete*. Elsevier Press.

Nixon, J.F., 1991. Discrete Ice Lens Theory for Frost Heave in Soils. *Canadian Geotechnical Journal*, Vol. 28, pp. 843–859.

Ohtsubo, M., Takayama, M., and Egashira, K., 1983. Relationships of Consistency Limits and Activity to Some Physical and Chemical Properties of Ariake Marine Clays. *Soils and Foundations*, Vol. 23, No. 1, pp. 38–46.

Rieke, R.D., Vinson, T.S., and Mageau, D.W., 1983. The Role of Specific Surface Area and Related Index Properties in the Frost Heave Susceptibility of Soils. *Proceedings of the 4th International Permafrost Conference*, pp. 1066–1071.

Ross, G.J., 1978. Relationships of Specific Surface Area and Clay Content to Shrink-Swell Potential of Soils Having Different Clay Mineralogic Compositions. *Canadian Journal of Soil Science*, Vol. 58, pp. 159–166.

Sridharan, A., Rao, S.M., and Dwarkanath, H.N., 1992. Dispersive Behavior of Nonswelling Clays. *Geotechnical Testing Journal*, Vol. 15, No. 4, pp. 380–387.

Smith, C.W., Hadas, A., Dan, J., and Koyumdjisky, H., 1985. Shrinkage and Atterberg Limits Relation to Other Properties of Principle Soil Types in Israel. *Geoderma*, Vol. 35, pp. 47–65.

Spangler, M.G., and Handy, R.L., 1982. *Soil Engineering*, 4th ed., Harper & Row.

ADDITIONAL READING

Akin, I.D., and Likos, W.J., 2014. Specific Surface Area of Clay Using Water Vapor and EGME Sorption Methods. *ASTM Geotechnical Testing Journal*, Vol. 37, No. 6, pp. 1–12.

Arnepalli, D.N., Shanthakumar, S., Hanumantha Rao, B., and Singh, D.N., 2008. Comparison of Methods for Determining Specific-Surface Area of Fine-Grained Soils. *Geotechnical and Geological Engineering*, Vol. 26, pp. 121–132.

deJong, E., 1999. Comparison of Three Methods of Measuring Surface Area of Soils. *Canadian Journal of Soil Science*, Vol. 79, pp. 345–351.

Dolinar, B., Misic, M., and Trauner, L., 2007. Correlation Between Surface Area and Atterberg Limits of Fine-Grained Soils. *Clays and Clay Minerals*, Vol. 55, No. 5, pp. 519–523.

Dyal, R.S., and Hendricks, S.B., 1950. Total Surface of Clays in Polar Liquids as a Characteristic Index. *Soil Science*, Vol. 69, pp. 421–432.

Eltantawy, I.M., and Arnold, P.W., 1973. Reappraisal of Ethylene Glycol Mono-Ethyl Ether (EGME) Method for Surface Area Estimations of Clays. *Soil Science Society of America Journal*, Vol. 24, No. 2, pp. 232–238.

Khorshidi, M., Lu, N., Akin, I.D., and Likos, W.J., 2017. Intrinsic Relationship Between Specific Surface Area and Soil Water Retention. *Journal of Geotechnical and Geoenvironmental Engineering, ASCE*, Vol. 143, No. 1, p. 11.

Lutenegger, A.J., and Cerato, A., 2001. Surface Area and Engineering Properties of Fine-Grained Soils. *Proceedings of the 15th International Conference on Soil Mechanics and Foundation Engineering*, Vol. 1, pp. 603–606.

Macht, F., Eusterhues, K., Pronk, G.J., and Totsche, K.U., 2011. Specific Surface Area of Clay Minerals: Comparison Between Atomic Force Microscopy Measurements and Bulk-Gas (N_2) and –Liquid (EGME) Adsorption Methods. *Applied Clay Science*, Vol. 53, pp. 20–26.

Muhunthan, B., 1991. Liquid Limit and Surface Area of Clays. *Geotechnique*, Vol. 41, No. 1, pp. 135–138.

Rai Puri, B., and Murari, K., 1963. Studies in Surface Area Measurements of Soils: 1. Comparison of Different Methods. *Soil Science*, Vol. 96, pp. 331–335.

Spagnoli, G., and Shimobe, S., 2021. A Statistical Reappraisal of the Relationship Between Liquid Limit and Specific Surface Area, Cation Exchange Capacity and Activity of Clays. *Journal of Rock Mechanics and Geotechnical Engineering*.

Sridharan, A., and Rao, G.V., 1972. Surface Area Determination of Clays. *Geotechnical Engineering*, Technical Note, Vol. 3, pp. 127–132.

Tiller, K.G., and Smith, L.H., 1990. Limitations of EGME Retention to Estimate the Surface Area of Soils. *Australian Journal of Soil Research*, Vol. 28, pp. 1–26.

Yukselen-Aksoy, Y., and Kaya, A., 2010. Method Dependency of Relationships Between Specific Surface Area and Soil Physicochemical Properties. *Applied Clay Science*, Vol. 50, pp. 182–190.

Chapter 10

Methylene Blue Test

INTRODUCTION

The Methylene Blue Test provides a method to give a measurement of the adsorption of methylene blue dye by clay. The test procedure gives the Methylene Blue Index (MBI) of the soil. The test is a simple indicator test that has been shown to be very useful in estimating other surface properties of fine-grained soils, such as Specific Surface Area (SSA) and Cation Exchange Capacity (CEC) when these laboratory tests are not readily available. The test method was developed in France in the 1970s (Lan 1977) for evaluating qualities of clay used for manufacturing ceramics, but it has grown in popularity in geotechnical engineering around the world as a means of evaluating the presence of smectite (montmorillonitic) clays in concrete aggregates. The test has now been used extensively in geotechnical studies, especially for fine-grained soils. The testing time takes between 15 and 60 minutes to complete, depending on the type and amount of clay.

STANDARD

ASTM C837 *Standard Test Method for Methylene Blue Index of Clay*. The procedure is also standardized in France by the Association Française de Normalization (AFNOR) under Norme Française NF P 94-068 Measure de le Quantité et de l'Acitité de le Fraction Argilluse.

EQUIPMENT

Baroid No. 987 filter paper (or Whatman No. 1)
Methylene Blue ($C_{16}H_{18}N_3SCl$) dye powder
Ceramic mixing bowls
Small mixer
Electronic scale with a minimum capacity of 250 gm and a resolution of 0.01 gm

DOI: 10.1201/9781003263289-11

pH meter
Convection oven capable of maintaining $110° \pm 5°C$
Ceramic or aluminum moisture tins
25 mL graduated cylinder
600 mL glass beaker
Ceramic or aluminum moisture tares
Spatula
Distilled water
Medicine dropper
Glass stirring rod
0.1N sulfuric acid
1000 mL glass containers (Erlenmeyer flask)

PREPARATION

Prepare methylene blue dye solution by placing 10.0 gm of methylene blue dye powder in a glass container and adding 1L of distilled water.

Prepare a solution of 0.1N sulfuric acid by adding 6.9 mL of concentrated sulfuric acid to 250 mL of distilled water. (Note: Always add the acid to the water. Pour slowly, mixing continuously.)

PROCEDURE

1. Select a representative sample of about 100 gm of air-dried soil passing the No. 40 sieve.
2. Place the sample in a small moisture tare and place in a convection oven at 110°C for 24 h.
3. After oven drying, place a specimen of 2.0 gm into the bottom of a clean dry 600 mL glass beaker.
4. Add 300 mL of distilled water to the beaker and stir gently until the soil is uniformly dispersed.
5. Determine the pH of the suspension using the pH meter. Add sufficient sulfuric acid to bring the pH within the range from 2.5 to 3.8.
6. Continue stirring while the pH is being adjusted and continue stirring for 10–15 min after the last addition of acid.
7. Fill the 25 mL graduated cylinder with the methylene blue solution and add 5 mL of the solution to the soil–water suspension. Stir for 1–2 min.
8. Remove a drop of the suspension using a dropper and place on the edge of the filter paper.
9. Observe the appearance of the drop on the filter paper. The drop should produce a dark blue stain surrounded by a colorless wet area. The end point of the test is indicated by the formation of a light blue halo around the drop.

10. Continue adding the methylene blue solution to the suspension in 1.0 mL increments and placing drops at a new location onto the filter paper until the end-point is reached. (Note: For more precise determinations, increments of 0.5 mL methylene blue solution may be used.)
11. At the end of the test, properly dispose of the materials in acceptable containers.

CALCULATIONS

Calculate the Methylene Blue Index (MBI) as:

$$MBI = (E \times V)/M \times 100 \tag{10.1}$$

where:
  E = milliequivalents of methylene blue per milliliter (Note: 1 ml = 0.01 meq)
  V = milliliters of methylene blue solution required for the titration
  M = dry mass of soil

If 2.0 gm of dry soil are used and the methylene blue titrating solution is 0.01N, Equation 10.1 simplifies to:

$$MBI = 0.01V/2 \times 100 = 0.5V \tag{10.2}$$

The Specific Surface Area (SSA) may be estimated from:

$$SSA = (1/319.87)(1/200)(0.5N)(A_V A_{MB}(1/10)) \tag{10.3}$$

where:
  SSA = Specific Surface Area (m²/gm)
  N = number of MB increments added to the soil suspension
  A_V = Avagadro's number (6.02×10^{23} mol)
  A_{BM} = area covered by 1 MB molecule (assumed = 130Å²)

EXAMPLE

A 2.00 gm sample of clay from Saint Herblain, France, gave the following results:
  Blue halo appeared after 5.5 mL of MB solution using 0.01N solution.
  From Equation 10.2:

$$MBI = 0.5(5.5\,mL) = 2.75$$

If the oven-dry mass of the soil is larger than 2.0 gm, Equation 10.1 should be used to calculate the MBI. SSA may then be estimated using Equation 10.3.

DISCUSSION

The test procedure describe in this chapter uses sulfuric acid, which should be handled with care. The use of laboratory gloves and eye protection is recommended. The formation of a light blue halo inside the wet area around the blue stain indicates the presence of an excess amount of methylene blue that is no longer adsorbed by the clay minerals in the suspension. Some variations of the test use a larger mass of soil (up to 30 gm) to increase the accuracy of the test (e.g., Locat et al. 1984; Cokca & Birand 1993). A comparison between the ASTM and AFNOR test procedures was given by Chiappone et al. (2004) showing a difference of about 10% between the two methods.

Prior to the development of the ASTM standard test procedure described, the test results and MBI had been used to estimate SSA and CEC through empirical studies and had been used to detect the presence of swelling clay minerals (Hang & Brindley 1970; Higgs 1988; Cokca & Birand 1993). More recent studies have shown good relationships between MBI and other soil characteristics, such as free-swell index and Hygroscopic Water Content (Chiappone et al. 2004; Sivapullaiah et al. 2008; Yukselen & Kaya 2008; Turkoz & Tosun 2011). The MBI has also been used to evaluate the effectiveness of lime modification of clays, similar to using the Initial Consumption of Lime test (ICL) described in Chapter 18 (Cambi et al. 2012). Typical values of MBI for different soils and clay minerals are given in Table 10.1.

Table 10.1 Reported Values of MBI for Different Soils

MBI	Soil	Reference
2.9	Kaolinite	Cokca (2002)
50.0	Bentonite	Chiappone et al. (2004)
12.3	Claystone	Abayazeed and El-Hinnawi (2011)
24.1	Montmorillonite	Abayazeed and El-Hinnawi (2011)

REFERENCES

Abayazeed, S.D., and El-Hinnawi, E., 2011. Characterization of Egyptian Smectite Clay Deposits by Methylene Blue Adsorption. *American Journal of Applied Sciences*, Vol. 8, No. 12, pp. 1282–1286.

Cambi, C., Carrisi, S., and Comodi, P., 2012. Use of the Methylene Blue Stain Test to Evaluate the Efficiency of Lime Treatment on Selected Clayey Soils. *Journal of Geotechnical and Geoenvironmental Engineering, ASCE*, Vol. 138, No. 9, pp. 1147–1150.

Chiappone, A., Marello, S., Scavia, C., and Setti, M., 2004. Clay Mineral Characterization Through the Methylene Blue Test: Comparison with Other Experimental Techniques and Applications of the Method. *Canadian Geotechnical Journal*, Vol. 41, pp. 1168–1178.

Cokca, E., 2002. Relationship Between Methylene Blue Value, Initial Soil Suction and Swell Precent of Expansive Soils. *Turkish Journal of Engineering and Environmental Science*, Vol. 26, pp. 521–529.

Cokca, E., and Birand, A., 1993. Determination of Cation Exchange Capacity of Clayey Soils by the Methylene Blue Test. *ASTM Geotechnical Testing Journal*, Vol. 16, No. 4, pp. 518–524.

Hang, P.T., and Brindley, G.W., 1970. Methylene Blue Adsorption by Clay Minerals: Determination of Surface Areas and Cation Exchange Capacities. *Clays and Clay Minerals*, Vol. 18, pp. 203–212.

Higgs, N.B., 1988. Methylene Blue Adsorption as a Rapid and Economical Method of Detecting Smectite. *ASTM Geotechnical Testing Journal*, Vol. 11, No. 1, pp. 68–71.

Lan, T.N., 1977. A New Test for the Identification of Soils—Methylene Blue Test. *Bulletin Liason Laboratoire Ponts et Chassee*, Vol. 88, pp. 136–137.

Locat, J., Lefebvre, G., and Ballivy, G., 1984. Mineralogy, Chemistry and Physical Properties Interrelationships of Some Sensitive Clays from Eastern Canada. *Canadian Geotechnical Journal*, Vol. 21, pp. 530–550.

Sivapullaiah, P.V., Prasad, B.G., and Allam, M.M., 2008. Methylene Blue Surface Area Method the Correlate with Specific Soil Properties. *ASTM Geotechnical Testing Journal*, Vol. 31, No. 6, pp. 503–512.

Turkoz, M., and Tosun, H., 2011. The Use of Methylene Blue Test for Predicting Swell Parameters of Natural Clay Soils. *Scientific Research and Essays*, Vol. 6, No., 8, pp. 1780–1792.

Yukselen, Y., and Kaya, A., 2008. Suitability of the Methylene Blue Test for Surface Area, Cation Exchange Capacity and Swell Potential Determination of Clayey Soils. *Engineering Geology*, Vol. 102, pp. 38–45.

ADDITIONAL READING

Kahr, G., and Madsen, F.T., 1995. Determination of the Cation Exchange Capacity and the Surface Area of Bentonite, Illite and Kaolinite by Methylene Blue Adsorption. *Applied Clay Science*, Vol. 9, pp. 327–336.

Kipling, J.J., and Wilson, R.B., 1960. Adsorption of Methylene Blue in the Determination of Surface Area. *Journal of Applied Chemistry*, Vol. 10, pp. 109–113.

Nevins, M.J., and Weintritt, D.J., 1967. Determination of Cation Exchange Capacity by Methylene Blue Adsorption. *Ceramics Bulletin*, Vol. 46, No. 6, pp. 587–592.

Phelps, G., and Harris, D., 1967. Specific Surface and Dry Strength by Methylene Blue Adsorption. *American Ceramic Society Bulletin*, Vol. 47, pp. 1146–1150.

Xidakis, G.S., and Smalley, I.J., 1980. Looking for Expansive Minerals in Expansive Soils: Experiments with Dye Adsorption Using Methylene Blue. *Proceedings of the 3rd Australia-New Zealand Conference on Geomechanics*, pp. 1203–1206.

Yukselen, Y., and Kaya, A., 2006. Comparison of Methods for Determining Specific Surface Area of Soils. *Journal of Geotechnical and Geoenvironmental Engineering, ASCE*, Vol. 132, No. 7, pp. 931–936.

Chapter 11

Pore Fluid Salinity

INTRODUCTION

The behavior of some fine-grained soils may be influenced considerably by the amount of soluble salts present in the pore fluid. The amount of soluble salts is often referred to as *Salinity*, which is the total amount of dissolved salts contained in 1 liter of water. Salinity is also referred to as *total dissolved solids* (TDS) or *soluble salts*. In soils, the salts are dissolved in the pore fluid between solid particles. Sea water has a salinity of 35 gm/L and contains sodium and potassium chloride along with other cations. The salinity of the pore fluid may help explain differences in behavior within a soil profile of a deposit. The determination of salt concentration may be especially important for investigations involving marine deposits and near coastal soils. Salts may be leached from marine deposits by infiltrating fresh water after deposition, which can alter their behavior considerably. The amount of Salinity in offshore marine deposits may also vary by geographic location as the availability of different salts in sea water may change.

Many investigations beginning in the 1960s have shown that the amount of salt in the pore fluid can influence a number of different characteristics of fine-grained soils, including Atterberg Limits (Arasan & Yetimoglu 2008; Pathak & Pathak 2016; Abu-Zeid & El-Aal 2017; Kumar et al. 2021), Activity (Ohtsubo et al. 1983; 2000), residual shear strength (Ramiah et al. 1970; Moore 1991; Di Maio & Fenelli 1994; Tiwari et al. 2005), compressibility (Mesri & Olsen 1971; Yukselen-Aksoy et al. 2008) and swelling (Warkentin & Schofield 1962; Biglari & Yasrobi 2005; Shirazi et al. 2011; Elmashad & Ata 2016). Studies of marine clay deposits in Canada, Sweden, Norway, Japan and Thailand have shown that the Salinity may vary from as little as 1 gm/L to as much as 25 gm/L. Investigations of these soils and others have shown that marine clays with low pore fluid salt concentrations generally have very high Sensitivity (often > 100), but the same deposits with high pore fluid salt concentrations have low Sensitivity (< 5) (e.g., Bjerrum 1954; Osterman 1965; Moum et al. 1971; Andersson-Skold et al. 2005; He et al. 2015).

DOI: 10.1201/9781003263289-12

Two methods of determining Pore Fluid Salinity are discussed in this chapter: 1) Salinity by refractometer; and 2) Salinity by electrical conductivity. Both methods require extraction of the pore fluid from the soil sample, which may be performed either by using a soil press or by centrifuge. Both methods are considered indirect.

PORE FLUID SALINITY BY REFRACTOMETER

The test procedure for determining Pore Fluid Salinity using a handheld refractometer is simple and involves first collecting pore fluid by squeezing the soil in a press and then using an inexpensive handheld light refractometer to obtain either a direct reading of Salinity in parts per thousand (ppt) or the Refraction Index depending on the model. Figure 11.1 shows a photo of a handheld refractometer. The current ASTM standard is based in part on work by Chaney et al. (1983). This method is based on the principle that the salt content in the pore fluid changes the angle of light refraction.

Standard

ASTM D4542 *Standard Test Method for Pore Water Extraction and Determination of the Soluble Salt Content of Soils by Refractometer*

Equipment

Refractometer (direct read scale in parts per thousand [ppt] or index of refraction)

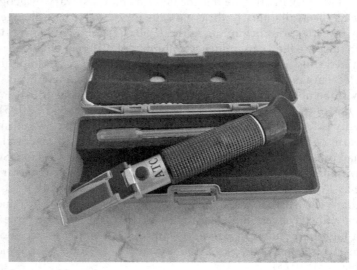

Figure 11.1 Handheld refractometer

Soil press (Note: A suitable soil press is shown in Figure 11.2; however, similar designs may be used, e.g., Torrance 1976).
Hydraulic press
25 mL plastic syringe
Electronic scale with a minimum capacity of 250 gm and a resolution of 0.01 gm
Filter paper (5–10 µm and 0.45 µm)
Refrigerator capable of maintaining constant temperature of 1–5°C
Micro-syringe filter holder
100 mL plastic or glass bottle with lid
Distilled water
Alcohol
Dilute hydrochloric acid (1:10)

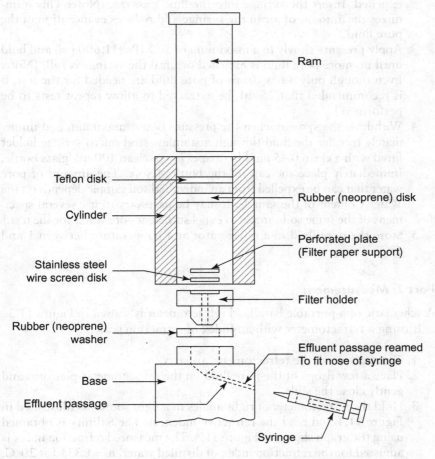

Figure 11.2 Soil press for extracting soil pore fluid (after Manheim 1966).

Preparation

Wash all parts of the soil press and stainless-steel micro-syringe holder thoroughly with distilled water and dry. Dry by a method that will not contaminate the parts, such as compressed air, oven or air drying, or rinsing with alcohol and air drying.

Procedure

Part 1: Extracting pore fluid

1. Select a representative sample of soil at natural Water Content of about 50–80 gm and place the soil into the soil press on top of a sheet of 5–10 μm filter paper to fit the press.
2. Apply pressure to the press until the first drops of pore water are expelled. Insert the syringe into the fluid passage. (Note: This minimizes the amount of air in the syringe and reduces evaporation of the pore fluid.)
3. Apply pressure slowly to a maximum of 80 MPa (11,600 psi) and hold until no more pore fluid is expelled or until the syringe is full. (Note: Even though only a few drops of pore fluid are needed for the test, it is recommended that 25 mL be extracted to allow repeat tests to be performed.)
4. Withdraw the syringe when the pressure is at a maximum and immediately transfer the fluid through a stainless-steel micro-syringe holder fitted with a clean 0.45 μm filter paper into a clean 100 mL glass bottle. Immediately place the cap on the bottle. (Note: The amount of pore water that can be expelled from an individual soil sample depends on the Water Content of the sample. It may be necessary to use several specimens of the same soil sample to expel sufficient pore water for the test.)
5. Store the pore fluid in a refrigerator at a temperature between 1 and 5°C.

Part 2: Measurement

A schematic of a portable handheld refractometer is shown in Figure 11.3. If using a refractometer with an index of refraction scale:

1. Carefully wash the refractometer and dry.
2. Place a few drops of the pore fluid on the refractometer platform and gently close the slide.
3. Hold the refractometer at right angles to a light source as illustrated in Figure 11.4 and read the refractive index, n. The Salinity is obtained using the graph shown in Figure 11.5. The measured refraction index is adjusted for the refraction index of distilled water, $n_0 = 1.3330$ at 20°C.

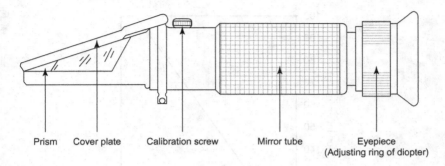

Prism Cover plate Calibration screw Mirror tube Eyepiece
(Adjusting ring of diopter)

Figure 11.3 Schematic of handheld refractometer.

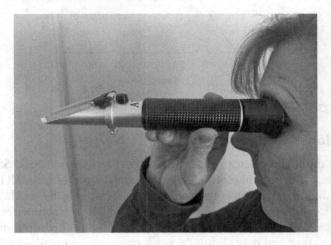

Figure 11.4 Holding refractometer to obtain Salinity reading.

If using a refractometer with a ppt (= gm/L) scale:

1. Carefully wash the refractometer and dry as previously noted.
2. Place one or two drops of the pore fluid into the semicircle of the white plastic area which is held firmly against the glass platform. Allow the fluid to escape under the white plastic area.
3. Hold the refractometer at right angles to a light source and read the Salinity to the nearest whole number. There should be a distinct black/white boundary. Read where the bottom of the refractometer hairline touches the beginning of the black boundary.

Calculations

There are no calculations for Salinity. Measurements are taken directly from the refractometer.

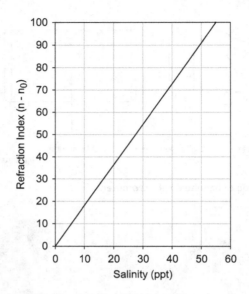

Figure 11.5 Graph of Salinity vs. adjusted refraction index.

Discussion

A small handheld refractometer generally cost less than around $50 and is easy to use, clean and maintain. There are now a number of small handheld portable digital refractometers available that can also be used. Many of these devices are more expensive than a handheld refractometer (around $300) but are easy to use. They provide a direct digital reading of Salinity and temperature and may eliminate any human error in reading a handheld refractometer. A typical example is shown in Figure 11.6.

PORE FLUID SALINITY BY ELECTRICAL CONDUCTIVITY

This method is based on the fact that the ability of electrical current to flow through the soil pore fluid depends on the concentration of salts present. As the concentration of salts increases, the electrical conductivity also increases. The method requires collection of the pore fluid which is then placed in a nonconductive container or "cell". A laboratory benchtop or portable conductivity meter, such as is shown in Figure 11.7, is then used to obtain a direct measurement of the electrical conductivity. The Salinity is obtained by using a calibration chart developed for that cell using fluids with different known amounts of salt. The method described in this section is based in part on the methods suggested by Bower and Wilcox 1965; Martin 1970; Janzen 1993; and Roades 1996. The measurement of Pore Fluid Salinity often can be conducted using the same pore fluid used to

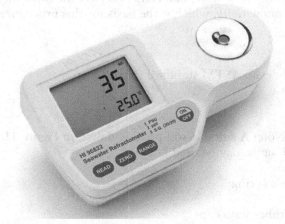

Figure 11.6 Handheld digital electronic refractometer.

Figure 11.7 Portable conductivity meter.

measure pH. Many benchtop pH meters also have the capability of measuring electrical conductivity, which is the basis for this procedure.

Standard

There currently is no ASTM standard for this test.

Equipment

Soil press (Note: A suitable soil press is shown in Figure 11.2; however, similar designs may be used.)
Hydraulic press
25 mL plastic syringe
Centrifuge
Centrifuge tubes with caps
Conductivity meter (alternating current)
Conductivity cell (glass beaker 150–250 ml)
Distilled water
Alcohol
Dilute hydrochloric acid (1:10)

Preparation

Calibration of the cell

The cell holding the pore fluid after extraction may be any nonconducting bottle or beaker provided that the cell with the same dimensions is used for all tests. The measurement of electrical conductivity is dependent on the cell. This means that the cell must be calibrated with distilled water containing known amounts of salt. Measure the conductivity of distilled water by placing approximately 100 mL of distilled water in 0 250 mL glass beaker. Insert the electrode (probe) from the conductivity meter and obtain the conductivity. Calibrate the meter using the manufacturer's operating instructions using standard saline solutions using known amounts of sea salt dissolved in 1L of distilled water, (e.g., of 5.0, 10.0, 15.0, 25.0 and 35 gm/L) using the procedure described below. (Note: Alternatively, standard conductivity solutions may be obtained from a chemical supply source to use for calibration.) An example of a calibration chart is shown in Figure 11.8.

Some researchers use a standard reference solution of 12.5 gm/L and use the normalized conductivity measurement (measured value/value of 12.5 gm/L) to create the calibration chart

Extracting pore fluid using a soil press

If using a soil press to extract pore fluid, follow the procedure described in Section 11.1.

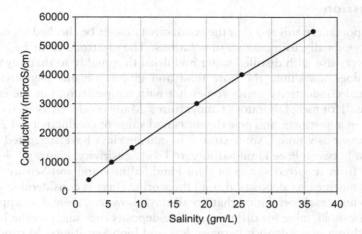

Figure 11.8 Typical conductivity meter calibration.

Extracting pore fluid using a centrifuge

The pore fluid may also be extracted by using a small benchtop centrifuge. Soil is placed in a clean dry centrifuge tube and a cap is placed on the tube. Several soil specimens may be centrifuged at the same time, provided that the load is balanced in the centrifuge. The use of the centrifuge eliminates the need to fabricate a soil press and generally only needs to operate for 15–20 min to provide supernatant for testing.

Procedure

1. Place the extracted pore fluid in the same conductivity cell used for calibration. Make sure that the cell is clean and dry. Make sure there is sufficient fluid so that the conductivity meter electrode will be fully immersed.
2. Follow the manufacturer's instructions for operating the conductivity meter and immerse the probe into the solution and read the conductivity from the meter.
3. Record the conductivity of the pore fluid. (Many conductivity meters provide a measure of conductivity in μS/cm or mS/cm (1mS/cm = 1000μS/cm.)
4. Use the calibration chart developed for the cell and meter and convert the conductivity measurement to Salinity and report the Salinity to the nearest 0.1 gm/L.

Calculations

There are no calculations with this procedure. The Salinity is obtained from the cell calibration chart using the measured conductivity.

Discussion

It is important in this test that the conductivity meter be checked for calibration periodically using standard solutions. The electrode must be cleaned after every use with distilled water and dried thoroughly so that any excess water does not dilute the pore fluid and give erroneous measurements. Electrical conductivity varies somewhat with temperature, but the error is very small for most laboratory temperatures. Many conductivity meters have a built-in adjustment and give the measured value of conductivity at 25°C.

As previously noted, some studies in marine clays have suggested a correlation between Pore Fluid Salinity and soil Sensitivity. Figure 11.9 shows results from reported values of Pore Fluid Salinity and soil Sensitivity for several marine clay deposits around the world. There is considerable scatter and no universal correlation that fits all clays; however, there does appear to be a threshold value for different marine deposits that suggests the boundary within a given deposit between low and high Sensitivity. Marine clays with Salinity less than about 1 gm/L should be investigated further as they may be indicative of highly sensitive material. Similar results have been presented by Duhaime et al. (2013) for other marine clays in Canada.

Electrical conductivity of intact or remolded soils is related in part to Pore Fluid Salinity as well as soil composition, e.g., mineralogy, CEC and clay content. A variety of in situ electrical probes have been used for mapping subsurface conditions at some sites. The author has also used the test procedure described in Section 11.2 to determine the combined electrical conductivity of soil:water suspension. This measurement provides an indication of the electrical conductivity that is related to both the soil composition and pore fluid composition. In this procedure, a 1:1 soil:water suspension is

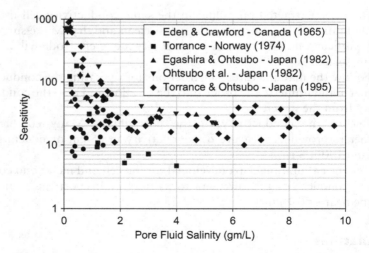

Figure 11.9 Trend between Pore Fluid Salinity and Sensitivity for several marine clays.

prepared by adding equal amounts of distilled water and air-dried soil to a glass beaker and stirring for 1 min with a glass stirring rod. The electrode from the conductivity meter is then immersed in the suspension and the electrical conductivity is read.

REFERENCES

Abu-Zeid, M.M., and El-Aal, A.K., 2017. Effect of Salinity of Groundwater on the Geotechnical Properties of Some Egyptian Clay. *Egyptian Journal of Petroleum*, Vol. 26, pp. 643–648.

Andersson-Skold, Y., Torrance, J.K., Lind, B., Oden, K., Stevens, R.L., and Rankka, K., 2005. Quick Clay – A Case Study of Chemical Perspective in Southwest Sweden. *Engineering Geology*, Vol. 82, pp. 107–118.

Arasan, S., and Yetimoglu, T., 2008. Effect of Inorganic Salt solutions on the Consistency Limits of Two Clays. *Turkish Journal of Engineering and Environmental Science*, Vol. 32, No. 2, pp. 107–115.

Biglari, M., and Yasrobi, S.S., 2005. Effects of NaCl Solution on Free Swelling Behavior of Compacted Clays. *Proceedings of the International Conference on Problematic Soils*, 8 pp.

Bjerrum, L., 1954. Geotechnical Properties of Norwegian Marine Clays. *Geotechnique*, Vol. 4, pp. 1–69.

Bower, C.A., and Wilcox, L.V., 1965. Soluble Salts. In *Methods of Soil Analysis: Part 2*, pp. 933–951. American Society of Agronomy.

Chaney, R.C., Slonim, S.M., and Slonim, S.S., 1983. Suggested Method for Determination of the Soluble Salt Content of Soils by Refractometer. *Geotechnical Testing Journal, ASTM*, Vol. 6, No. 2, pp. 93–95.

Di Maio, C., and Fenelli, G.B., 1994. Residual Strength of Kaolin and Bentonite: The Influence of their Constituent Pore Fluid. *Geotechnique*, Vol. 44, No. 4, pp. 217–226.

Duhaime, F., Benabdallah, E.M., and Chapuis, R.P., 2013. The Lachenaie Clay Deposit: Some Geotechnical and Geotechnical Properties in Relation to the Salt-Leaching Process. *Canadian Geotechnical Journal*, Vol. 50, pp. 311–325.

Elmasad, M.E., and Ata, A.A., 2016. Effect of Sea Water on Consistency, Infiltration Rate and Swelling Characteristics of Montmorillonite Clay. *Housing and Building Research Center Journal*, Vol. 12, pp. 175–180.

He, P., Ohtsubo, M., Higashi, T., and Kanayama, M., 2015. Sensitivity of Sal-Leached Clay Sediments in the Ariake Bay Area, Japan. *Marine Georesources and Geotechnology*, Vol. 33, No.5, pp. 429–436.

Janzen, H.H., 1993. Soluble Salts. Chapter 18. In M.R. Carter (Ed.), *Soil Sampling and Methods of Analysis*, pp. 161–166. Canadian Society of Soil Science.

Kumar, S., Garg, P., and Singh, A., 2021. A Study on the Influence of Sodium Salt Solution on the Atterberg Limits and Swelling of Bentonite Clay. *Journal of the University of Shanghai for Science and Technology*, Vol. 23, No. 10, pp. 332–341.

Manheim, F.T., 1966. A Hydraulic Squeezer for Obtaining Interstitial Water from Consolidated and Unconsolidated Sediments. *U.S. Geological Survey Professional Paper 550-e*, pp. 256–261.

Martin, R.T., 1970. Suggested Method of Test for Soluble Salts in Soil. *ASTM STP*, Vol. 479, pp. 288–290.

Mesri, G., and Olson, R., 1971. Consolidation Characteristics of Montmorillonite. *Geotechnique*, Vol. 21, No. 4, pp. 341–353.

Moore, R., 1991. The Chemical and Mineralogical Controls Upon the Residual Strength of Pure and Natural Clays. *Geotechnique*, Vol. 41, No. 1, pp. 35–47.

Moum, J., Loken, T., and Torrance, J.K., 1971. A Geotechnical Investigation of the Sensitivity of a Normally Consolidated Clay from Drammen, Norway. *Geotechnique*, Vol. 21, pp. 329–340.

Ohtsubo, M., Egashira, K., Kuomoto, T., and Bergado, D.T., 2000. Mineralogy and Chemistry, and Their Correlation with the Geotechnical Index Properties of Bangkok Clay: Comparison with Ariake Clay. *Soils and Foundations*, Vol. 40, No. 1, pp. 11–21.

Ohtsubo, M., Takayama, M., and Egashira, K., 1983. Relationships of Consistency Limits and Activity to Some Physical and Chemical Properties of Ariake Clays. *Soils and Foundations*, Vol. 23, No. 1, pp. 38–46.

Osterman, J., 1965. Studies on the Properties and Formation of Quick Clays. *Swedish Geotechnical Institute*, Vol. 8.

Pathak, Y., and Pathak, A., 2016. Effect of Saline Water on Geotechnical Properties of Soil. *International Journal of Innovative Research in Science, Engineering and Technology*, Vol. 5, No. 9, pp. 16181–16187.

Ramiah, B.K., Dayalu, N.K., and Purushothamaraj, P., 1970. Influence of Chemicals on Residual Strength of Silty Clay. *Soils and Foundations*, Vol. 10, No. 1, pp. 25–36.

Roades, J.D., 1996. Salinity: Electrical Conductivity and Total Dissolved Solids. In *Methods of Soil Analysis: Part 3, Chemical Methods, Soil Science Society of America*, pp. 417–436.

Shirazi, S.M., Wiwat, S., Kazama, H., Kuwano, J., and Shaaban, M.G., 2011. Salinity Effect on Swelling Characteristics of Compacted Bentonite. *Environmental Protection Engineering*, Vol. 37, No. 2, pp. 65–74.

Tiwari, B., Tuladhar, G.R., and Marui, H., 2005. Variation in Residual Shear Strength of the Soil with Salinity of Pore Fluid. *Journal of Geotechnical and Geoenvironmental Engineering, ASCE*, Vol. 131, No. 12, pp. 1445–1456.

Torrance, J.K., 1976. Pore Water Extraction and the Effect of Sample Storage on the Pore Water Chemistry of Leda Clay. *ASTM STP*, Vol. 598, pp. 147–157.

Warkentin, B.P., and Schofield, R.K., 1962. Swelling Pressures of Na-Montmorillonite. *Soil Science Society of America Proceedings*, Vol. 13, No. 1, pp. 98–105.

Yukselen-Aksoy, Y., Kaya, A., and Oren, A.H., 2008. Seawater Effect on Consistency Limits and Compressibility Characteristics of Clays. *Engineering Geology*, Vol. 102, pp. 54–61.

ADDITIONAL READING

Ajmera, B., Tiwari, B., and Ostrova, F., 2018. Influence of Salinity of Pore Fluid on the Undrained Shear Strength of Clays. *Proceedings of the International Foundations Congress and Equipment Exposition*.

Carson, M.A., 1981. Influence of Pore Fluid Salinity of Instability of Sensitive Clays: A New Approach to an Old Problem. *Earth Surface Processes and Landforms*, Vol. 6, pp. 499–515.

He, P., Ohtsubo, M., Higahi, T., and Kanayama, M., 2015. Sensitivity of Salt-Leached Clay Sediments in the Ariake Bay Area, Japan. *Marine Georesources and Geotechnology*, Vol. 33, pp. 429–436.

Eden, W.J., and Crawford, C.B., 1965. Geotechnical Properties of Leda Clay in the Ottawa Area. *Proceedings of the 6th International Conference on Soil Mechanics and Foundation Engineering*, Vol. 1, pp. 22–27.

Egashira, K., and Ohtsubo, M., 1983. Smectite in Marine Quick-Clays of Japan. *Clays and Clay Minerals*, Vol. 30, No. 4, pp. 275–280.

Liu, J., Afroz, M., and Ahmad, A., 2021. Experimental Investigation of the Impact of Salinity on Champlain Sea Clay. *Marine Georesources and Geotechnology*, Vol. 39, No. 4, pp. 494–504.

Lodahl, M.R., and Sorensen, K.K., 2019. Effects of Pore Water Chemistry on the Unloading-Reloading Behavior of Reconstituted Clays. *Proceedings of the 17th European Conference on Soil Mechanics and Geotechnical Engineering*.

Long, M., Donohue, S., L'Heureux, J.S., O'Conner, P., Sauvin, G., Romoen, M., and Lecomte, I., 2012. Relationship Between Electrical Resistivity and Basic Geotechnical Parameters for Marine Clays. *Canadian Geotechnical Journal*, Vol. 49, pp. 1–11.

Ridlo, A., Ohtsubo, M., Higashi, T., Kanayama, M., and Tanaka, M., 2012. Effects of Pore Water Salinity on the Liquid Limit of Mexico City Clay and the Swelling Characteristics of its Constituent Minerals. *Clay Science*, Vol. 16, pp. 105–110.

Soberg, I.L., Ronning, J.S., Dalsegg, E., Hansen, L., Rokoengen, K., and Sandven, P., 2008. Resistivity Measurements as a Tool for Outlining Quick Clay Extents and Valley-Fill Stratigraphy: Feasibility Study from Buvika, Central Norway. *Canadian Geotechnical Journal*, Vol. 45, pp. 210–225.

Torrance, J.K., 1974. A Laboratory Investigation of the Effect of Leaching on the Compressibility and Shear Strength of Norwegian Marine Clays. *Geotechnique*, Vol. 24, No. 2, pp. 155–173.

Torrance, J.K., 1984. A Comparison of Marine Clays from Ariake Bay, Japan and the South Nation River Landslide Site, Canada. *Soils and Foundations*, Vol. 24. No. 2, pp. 75–81.

Torrance, J.K., and Ohtsubo, M., 1995. Ariake Bay Quick Clays: A Comparison with the General Model. *Soils and Foundations*, Vol. 35, No. 1, pp. 11–19.

Part II

Determination of
Behavioral Characteristics

Consistency of Fine-Grained Soils

Atterberg Liquid and Plastic Limits

BACKGROUND: THE ATTERBERG CONCEPT

The Atterberg Limits are used to describe the phase changes that a fine-grained soil passes through at different Water Contents. At very high Water Content, a mixture of soil and water behaves as a viscous liquid. As the Water Content is reduced, the soil takes on the characteristics of a plastic solid and then a semisolid; finally, at a sufficiently low enough Water Content, the soil becomes a brittle solid. The Water Contents where these changes in behavior occur are called the Atterberg Limits, after Albert Atterberg, a Swedish chemist and agricultural scientist, as illustrated in Figure 12.1.

The Liquid Limit is defined as the Water Content that separates liquid behavior from plastic solid behavior. The Plastic Limit is defined as the Water Content that separates plastic solid behavior from semisolid behavior. The Shrinkage Limit is defined as the Water Content that separates semisolid behavior from brittle solid behavior. These are ideal definitions but we need some simple tests to help us define these boundaries. Over the years, standardized tests have been developed for this and generally seem to provide very good results and are able to separate different soil behavior based on the tests. In this chapter, tests for Liquid Limit and Plastic Limit are described.

Soils will behave differently depending on both the absolute values of the Liquid, Plastic and Shrinkage Limits and on the relative difference between these values. The Atterberg Limits can be used to indicate relative soil stiffness in a general way by locating the natural Water Content of the soil relative to the positions of the Liquid and Plastic Limits, as shown in Figure 12.1. For example, if the Water Content of a soil is near the Liquid Limit, we may expect soft soil; if the Water Content is near the Plastic Limit, we may expect stiff soil; and if the Water Content is near the Shrinkage Limit, we may expect very stiff or hard soil.

DOI: 10.1201/9781003263289-14

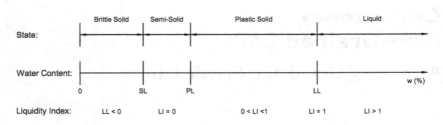

Figure 12.1 Conceptual model of Atterberg Limits.

The Atterberg Liquid and Plastic Limits are often used to help separate different fine-grained soils by their likely behavior. It should be remembered that these tests have arbitrary definitions based on standardized test procedures and are performed on remolded soils, but they have a long history of use and are therefore important to geotechnical engineering. While they do not necessarily reflect the true and accurate behavior of undisturbed soils, they do provide a method of comparing different fine-grained soils. In effect, the Atterberg Limits help separate different fine-grained soils by the way in which they interact with water.

Even though the Atterberg Limits are Water Contents where changes in soil behavior occur, they are normally reported as whole numbers and not percentages.

LIQUID LIMIT BY CASAGRANDE DROP CUP

Introduction

The Liquid Limit is defined as the Water Content above which the soil behaves as a viscous liquid. But in reality, the Liquid Limit was defined somewhat arbitrarily by Casagrande (1932) as the Water Content at which a specified size groove cut into a remolded soil–water paste placed in a shallow brass cup will flow closed when dropped 25 times a distance of 1 cm onto a hard rubber base. This definition may seem unscientific, but it is simple, actually works quite well, and has generally served the profession well. It has been used successfully in most cases for the classification of soils and in distinguishing (indirectly) differences in soil behavior. For the past 80 years, the Casagrande drop cup has been the most common method of determining Liquid Limit. Later in this chapter, an alternative method using the Fall Cone device is presented. A schematic of the Casagrande drop cup is shown in Figure 12.2.

Standard

ASTM D4318 *Standard Test Methods for Liquid Limit, Plastic Limit, and Plasticity Index of Soils*

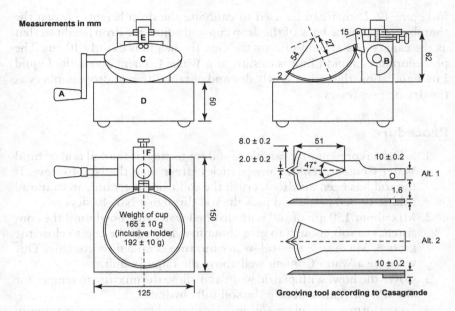

Figure 12.2 Schematic of Casagrande drop cup.

Equipment

> Casagrande drop cup apparatus
> Grooving tool
> Ceramic or plastic mixing bowls
> No. 40 sieve
> Ceramic mortar and rubber-tipped pestle
> Electronic scale with a minimum capacity of 250 gm and a resolution of
>   0.01 gm
> Convection oven capable of maintaining $110 \pm 5°C$
> Distilled water
> Plastic wash bottle
> Plastic wrap
> Spatula
> Ceramic or aluminum moisture tares

Calibration of Casagrande Drop Cup

The Casagrande drop cup is a simple device that uses a hand crank and
cam to raise and drop the cup onto a hard rubber base. The drop height
must be calibrated so that the test results are valid. Usually, the calibration is performed in between soil samples. The grooving tool that comes
with the drop cup and a handle is equipped with a 10 cm strip, shown

in Figure 12.2, that can be used to calibrate the drop height. Loosen the thumb screws on the back of the drop cup and adjust the drop height so that as the cam comes around to drop the cup, the height is exactly 10 cm. The procedure for Liquid Limit is to start at a Water Content above the Liquid Limit and allow the soil to slowly dry and obtain different drop numbers as the drying progresses.

Procedure

1. Select a representative sample of soil for testing. If the soil is at natural Water Content, remove any particles larger than the No. 40 sieve. If the soil has been air dried, crush the soil using a ceramic mortar and rubber-tipped pestle and pass the soil through a No. 40 sieve.
2. Mix about 150 gm of soil with distilled water in a bowl until the consistency is soft enough to give about five drops of the cup to close the groove. Make sure the soil–water mixture is mixed thoroughly. This will give a Water Content well above the Liquid Limit.
3. Cover the bowl with plastic wrap and allow the mixture to temper for about 24 h. This will help the soil fully hydrate.
4. To perform a test, place soil in a clean dry brass cup to a maximum depth of about 0.5 in (12.5 mm). Smooth the soil surface with the spatula.
5. Use the grooving tool to cut a groove perpendicular to the cup, as shown in Figure 12.3.

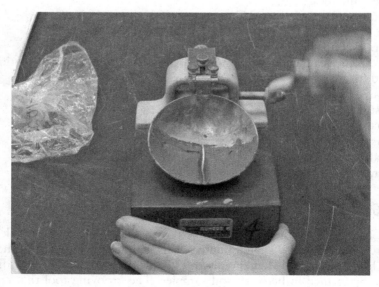

Figure 12.3 Photo of Casagrande Liquid Limit drop cup.

6. Turn the crank at a rate of about two drops per sec until the groove closes for a length of 0.5 in (12.5 mm). Record the number of drops. (Note: The goal of the test is to obtain 4 or 5 points at different Water Content, with drops ranging from about 10 to 35.)

7. Quickly remix the soil in the bowl and repeat the test until two consistent counts (± 1 drop) are obtained.

8. Obtain a Water Content sample of about 10–15 gm by using the spatula to cut across the closed groove. Place the sample in a moisture tin and obtain the Water Content using the convection oven method as described in Chapter 1.

9. Return the remaining soil to the bowl and clean the cup.

10. Allow the soil to dry slowly and repeat the procedure to obtain 4 or 5 separate determinations between 10 and 35 drops. (Note: If the soil dries too quickly and 4 or 5 measurements cannot be obtained, add more water to the soil, mix thoroughly and allow the soil to temper again.)

Calculations

For each sample collected, calculate the Water Content of the soil using:

$$W = (M_{WATER}/M_{DRY}) \times 100\% \qquad (12.1)$$

where:

$M_{WATER} = \text{mass of water} = (M_{WET} + M_{TARE}) - (M_{DRY} + M_{TARE})$

$M_{DRY} = \text{oven dry mass of soil} = (M_{DRY} + M_{TARE}) - M_{TARE}$

Make a semi-log plot of Water Content (y-axis) vs. number of drops (x-axis).

From the Flow curve, the Liquid Limit is determined as the Water Content corresponding to 25 drops of the cup.

Example

A sample of alluvial clay from Mississippi was tested to determine the Liquid Limit. The results are given in Table 12.1.

Table 12.1 Sample of Alluvial Clay

No. of drops	Water Content (%)
6	70.1
13	62.5
24	58.6
36	54.8

From the plot of no. of drops vs. Water Content shown in Figure 12.4, the interpreted Liquid Limit is 58. Note that even though the Water Content for each individual measurement was calculated to the nearest 0.1%, the Liquid Limit is reported as an integer and is only reported as a number, not a percentage.

The straight line shown in Figure 12.4 is called the *flow curve* and the slope of the line is called the *flow index*. To obtain the flow index, we just need to determine the numerical difference in Water Content over one log cycle of drops, for example, between 6 and 60. In this case, the flow index is 77.8 − 50 = 27.8. The flow index actually describes how quickly the remolded undrained shear strength of a soil changes with decreasing or increasing Water Content. A high value indicates that a large amount of water is needed to change the shear strength the same amount as a soil that has a low value, which indicates that only a small amount of water would cause the same change in strength. Figure 12.5 shows the flow curve of three different soils. The flow curve will always be linear when plotted on a semi-log plot. The reason for this will be discussed later. Another parameter that may be useful is the Toughness Index, defined as P.I./flow index.

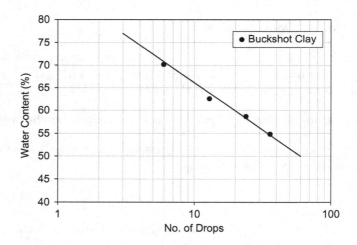

Figure 12.4 Liquid Limit flow curve.

Discussion

If the Water Content is normalized (divided by) by the Liquid Limit for each of the soils, the data for all three soils fall on a single trend line, shown in Figure 12.6. This is characteristic of most fine-grained soils. In some way, the Liquid Limit is a common denominator among different soils. This plot actually makes sense since the number of drops of the cups represents some

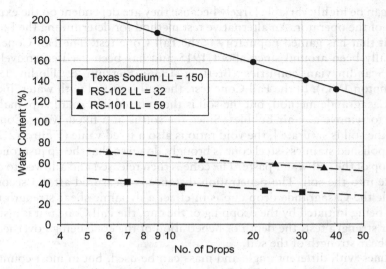

Figure 12.5 Typical results obtained for three clays from Casagrande drop cup.

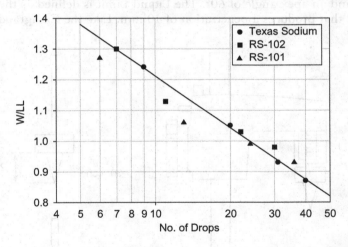

Figure 12.6 Normalized Liquid Limit results obtained from Casagrande drop cup.

measure of undrained shear strength, and a semi-log plot of Water Content versus remolded undrained shear strength is linear.

LIQUID LIMIT BY FALL CONE

Introduction

Even though the Casagrande drop cup has been in use for over 80 years, some engineers have argued that results from the Casagrande Liquid Limit cup can be highly variable largely because they are dependent on the experience of the operator. An alternative test method for determining the Liquid Limit that has gained popularity is the Fall Cone test. The Fall Cone has actually been around since about 1915 and has been used extensively in the Scandinavian countries (Bretting 1936; Bjerrum & Flodin 1960; Skempton 1985). In the Fall Cone test, the soil is mixed with water, like in the Casagrande method, but the soil is then placed in a small cup, making sure to remove any air bubbles. Since the soil is at a fixed Water Content and the soil is saturated, the void ratio is also a fixed value (Figure 12.7).

A polished stainless-steel cone is brought down so that the tip just touches the top of the soil/water paste. The cone is then released and allowed to penetrate into the soil. The penetration distance is measured after 5 seconds. While the Casagrande drop cup is in effect a dynamic shear strength test, flow being initiated by the dropping of the cup, the Fall Cone test is a static shear strength test, the depth of penetration of the cone being governed by the shear strength of the soil.

Cones with different angle and mass can be used, but in most countries the Liquid Limit from the Fall Cone is obtained using a cone with a mass of 60 gm and an apex angle of 60°. The Liquid Limit is defined as the Water Content that produces a penetration of 10 mm. Like the Casagrande drop

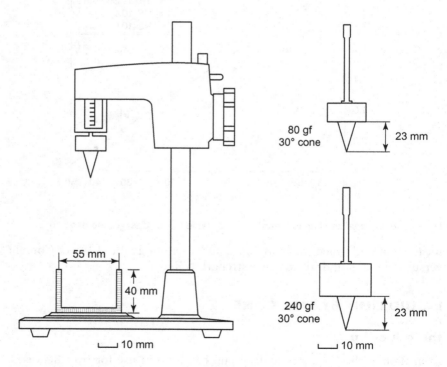

Figure 12.7 Schematic diagram of Fall Cone device.

cup method, the soil is initially mixed to a high Water Content and then several measurements of cone penetration are obtained on either side of 10 mm as the soil dries slowly in order to obtain the flow curve and the Liquid Limit.

Standard

There currently is no ASTM standard for the Fall Cone, although the process of developing a test standard has been initiated. The Fall Cone is standardized by the Standards International Organization (ISO) and is also standardized by the British Standard BS 1377, which uses a cone with an apex angle of 30° and a mass of 80 gm and defines the Liquid Limit as the Water Content which produces a penetration of 20 mm.

Equipment

Fall Cone device (Figure 12.8)
60 gm – 60° cone or 80 gm – 30° cone
Ceramic or plastic mixing bowls
No. 40 sieve
Ceramic mortar and rubber-tipped pestle
Electronic scale with a minimum capacity of 250 gm and a resolution of
    0.01 gm

Figure 12.8 Fall Cone device with different cones.

Convection oven capable of maintaining $110° \pm 5°C$

Distilled water

Plastic wrap

Spatula

Ceramic or aluminum moisture tares

Stop watch

Procedure

1. Select a representative sample of soil for testing. If the soil is at natural Water Content, remove any particles larger than the No. 40 sieve. If the soil has been air dried, crush the soil using a ceramic mortar and rubber-tipped pestle and pass the soil through a No. 40 sieve.
2. Mix about 150 gm of soil with distilled water in a bowl until the consistency is soft enough to give about five drops of the cup to close the groove. Make sure the soil–water mixture is mixed thoroughly. This will give a Water Content well above the Liquid Limit. Cover the bowl with plastic wrap and allow the mixture to temper for about 24 h. This will help the soil fully hydrate.
3. After tempering, place the mixture into a Fall Cone cup using a spatula. Make sure that all of the air has been removed and then level off the top of the cup to give a flat surface.
4. Place the cup onto the Fall Cone device and lower the cone so that the point just touches the surface of the soil. (Note: Make sure the cone is clean and dry.)
5. Depending on which Fall Cone device is being used, adjust the penetration depth gauge to zero, or take an initial reading.
6. Release the cone and start the stop watch. After 5 sec, secure the cone and obtain the penetration depth from the penetration gauge or obtain a final reading.
7. Collect a soil sample of about 20 gm from the middle of the cup and determine the Water Content using the convection oven method described in Chapter 1.
8. Allow the soil to slowly dry while periodically remixing to make sure the Water Content is uniform throughout.
9. Repeat Steps 6 and 7 so that four or five determinations of penetration and Water Content are obtained over a range of penetration. (Note: For a 60 gm – 60° cone this should be from about 25 mm to 5 mm; for an 80 gm – 30° cone, this should be about from 40 mm to 10 mm.)

Calculations

For each measurement of penetration, determine the Water Content. Make a semi-log plot of Water Content vs. penetration depth. This is the flow curve. The Liquid Limit is defined as the Water Content corresponding to a penetration depth of 10 mm for a 60 gm – 60° cone or 20 mm for an 80 gm – 30° cone.

Example

Fall Cone tests using a 60 gm – 60° cone were performed on the same soil previously tested using the Casagrande drop cup. The results are given in Table 12.2. Figure 12.9 shows the flow curve from these data. From the flow curve the Liquid Limit is interpreted as: L.L. = 61.

Figure 12.10 shows results of Fall Cone tests performed on the same soils as previously shown in Figure 12.5. The results look very similar as those obtained using the Casagrande drop cup. As with the Casagrande cup data, the Fall Cone data for these three soils have been normalized with respect to the Liquid Limit for each soil. The results are shown in Figure 12.11 and are similar to the normalized results obtained using the Casagrande cup previously shown in Figure 12.6.

Table 12.2 Fall Cone Tests

Trial no.	Penetration (mm)	Water Content (%)
1	15.0	67.0
2	12.2	63.6
3	8.0	57.7
4	5.0	50.0

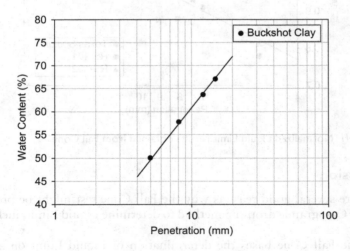

Figure 12.9 Flow curve from Fall Cone Liquid Limit test.

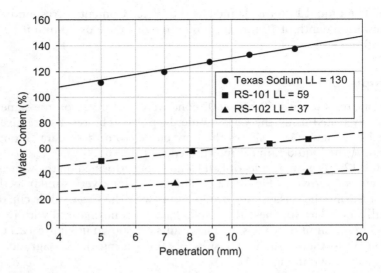

Figure 12.10 Typical results obtained for three soils from Fall Cone for Liquid Limit.

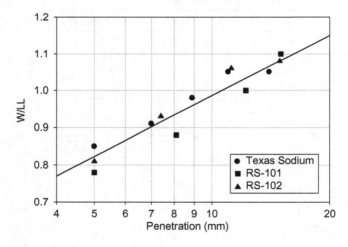

Figure 12.11 Normalized Liquid Limit results obtained from Fall Cone.

Discussion

There are several good reasons why the Fall Cone test might be preferred over the Casagrande drop cup method to determine Liquid Limit, including:

1. The Fall Cone bases the determination of Liquid Limit on a static rather than dynamic response of the soil.
2. The method is generally less operator dependent.

3. The method is generally less dependent on differences in equipment.
4. The test is simple and easy to perform.

In addition, the Fall Cone test has other applications that the geotechnical engineer may be interested in and which the Casagrande drop cup does not apply:

5. The test may be used on both undisturbed and remolded soils.
6. Different combinations of cone angle and mass may be used to test a wide range of soils.
7. The test provides a direct measurement of undrained shear strength.
8. The soil Sensitivity may be evaluated by first testing an undisturbed soil and then repeating the test on the same soil after being fully remolded.
9. The variation in remolded strength with Water Content may be determined by testing the same soil at different Water Content.
10. The test may also be used to estimate the Plastic Limit.
11. The test may also be used to estimate other soil properties.

Many studies have shown that the results of the Casagrande drop cup and Fall Cone give nearly the same results for Liquid Limit, at least for Liquid Limit values up to about LL = 120. An example is shown in Figure 12.12.

A true liquid has a shear strength of zero, but a remolded soil at a Water Content equal to the Liquid Limit, which is just at the boundary between

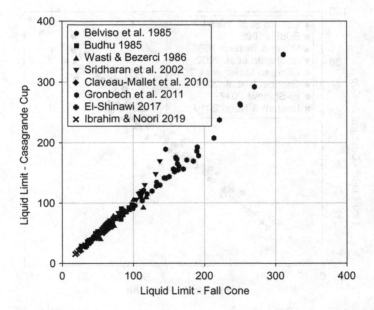

Figure 12.12 Comparison between Liquid Limit by Casagrande drop cup and Fall Cone.

a liquid and a plastic material and is not quite a liquid yet so it has a very small but measurable shear strength. The basis of using the Fall Cone to determine the Liquid Limit of a soil is that by definition, the Water Content obtained at a fixed penetration or rather providing a fixed undrained shear strength which is constant and independent of soil type.

This definition is somewhat arbitrary, but arguably no more arbitrary as that used by Casagrande, i.e., 25 blows of the drop cup. Despite the assumed idealized behavior which is taken for simplicity, soils possess a small, but measurable, undrained shear strength at a Water Content equal to the Casagrande Liquid Limit. However, existing data indicate that the undrained shear strength is not a constant at W = Liquid Limit, but varies slightly for different soils from about 1.2 to 2.2 kPa. Some of this variation may be attributed to the relative magnitude of the absolute value of s_u, and some variation may simply be related to test error (Figure 12.13).

Hansbo (1957) suggested that the shear strength of soil from a Fall Cone test could be determined from:

$$s_u = 9.8\big[(M_c)(K)\big]/d^2 \tag{12.2}$$

where:
  s_u = undrained shear strength (kPa)
  M_c = mass of the cone (gm)

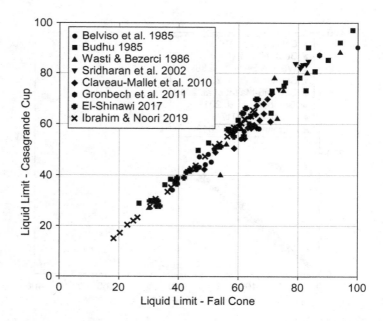

Figure 12.13 Comparison between Liquid Limit by Casagrande drop cup and Fall Cone.

K = a constant depending on cone apex angle (K = 0.8 for 30° cone; K = 0.27 for 60° cone)

D = depth of penetration (mm)

For a 60 gm – 60° cone, Equation 12.2 reduces to:

$$s_u = 9.8\big[(60)(0.27)\big]/d^2 \qquad\qquad (12.3)$$

and for a penetration of 10 mm

$$s_u = 1.6\,kPa$$

For an 80 gm – 30° cone, Equation 12.2 reduces to:

$$s_u = 9.8\big[(80)(0.8)\big]/d^2 \qquad\qquad (12.4)$$

and for a penetration of 20 mm

$$s_u = 1.6\,kPa$$

This means that using either cone results in obtaining the Water Content that corresponds to an undrained shear strength of 1.6 kPa. Therefore, the Liquid Limit is defined as the Water Content that gives an undrained shear strength of a remolded soil equal to 1.6 kPa.

According to Eq. 12.2, the Liquid Limit determined by both the 60 gm – 60° and an 80 gm – 30° cone should provide the same value. There have been only a few studies comparing the two methods; however, Leroueil and LeBihan (1996) and Farrell (1997) showed that the two methods give the same value of Liquid Limit over a range of Liquid Limit of about 20 to 180. Figure 12.14 shows some of the available results and results obtained by the author.

PLASTIC LIMIT BY ROLLING THREAD

Introduction

Traditionally, the Plastic Limit is determined using the thread method by rolling a soil–water mixture into a thread with a diameter of 3.2 mm (1/8 in) on a clean dry glass plate. At a Water Content equal to the Plastic Limit, the thread will just begin to crumble. If the thread crumbles before it gets to 3.2 mm (1/8 in), then the soil is too dry; if it gets to (3.2 mm (1/8 in) but doesn't crumble, the soil is too wet. The procedure requires considerable experience, and different technicians can easily report different values.

The thread method for determining Plastic Limit is a compliment to the Casagrande drop cup method for determining the Liquid Limit. The Plastic

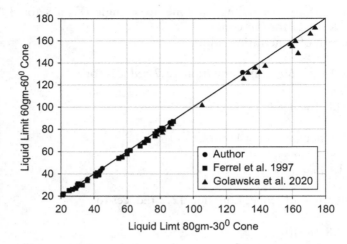

Figure 12.14 Comparison between Liquid Limit from 60gm 60° cone and 80gm 30° cone.

Limit and Liquid Limit are usually obtained together using the same mixture of soil + water. The test is an arbitrary procedure and really has very little scientific basis, but it has been in use for over 80 years and, for all of its potential problems, it actually gives reasonable results of soil behavior, provided everyone uses the same procedure. The Plastic Limit determination by the thread method tends to be more variable than the Liquid Limit determined using the Casagrande drop cup.

Standard

ASTM D4318 *Standard Test Methods for Liquid Limit, Plastic Limit, and Plasticity Index of Soils*

Equipment

Smooth glass plate approximately 30.5 cm × 30.5 cm (12 in × 12 in)
3.2 mm (1/8 in) stainless-steel reference rod
Ceramic or plastic mixing bowls
Spatula
Convection oven capable of maintaining 110° ± 5°C
Electronic scale with a minimum capacity of 250 gm and a resolution of
    0.01 gm

Procedure

1. At the beginning of the Liquid Limit test, separate about 25 gm of the tempered soil–water mixture before starting the Liquid Limit test

and set aside to dry. (Note: If the Plastic Limit test is being performed without a Liquid Limit test, select a representative sample of soil for testing. If the soil is at natural Water Content, remove any particles larger than the No. 40 sieve. If the soil has been air dried, crush the soil using a ceramic mortar and rubber-tipped pestle and pass the soil through a No. 40 sieve. Mix about 50 gm of soil with distilled water in a bowl until the consistency is soft enough to give about 10 drops of the cup to close the groove cut with the grooving tool. Make sure the soil–water mixture is mixed thoroughly. Cover the bowl and set the mixture aside to temper for 24 h.

2. Periodically, remix the soil with a spatula so that the Water Content is uniform throughout.

3. When the soil has dried to a "plastic" state, roll about 1/3 of the soil from the mix into an 1/8 in (3.2 mm) diameter thread on the glass plate using your hand. Use the stainless-steel rod as a reference for 1/8 in (3.2 mm), as shown in Figure 12.15.

4. Repeat Step 3 until the thread shows signs of crumbling when it reaches a diameter of 1/8 in (3.2 mm). If the soil crumbles before it can be rolled to a diameter of 1/8 in (3.2 mm), the soil is too dry; if the soil can be rolled to a diameter less than 1/8 in (3.2 mm), the soil is too wet. The goal is to have the thread just begin to crumble when it reaches a diameter of 1/8 in (3.2 mm).

5. Collect a sample of about 5–8 gm and determine the Water Content using the convection oven method described in Chapter 1.

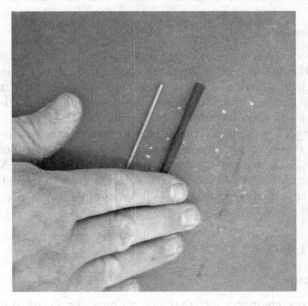

Figure 12.15 Technician rolling soil into a thread to determine Plastic Limit.

6. Repeat Steps 4 and 5 for the remaining 2/3 of the sample using 1/3 at a time. (Note: In between trials, cover the mixture so that it doesn't dry out further.)
7. Calculate the Plastic Limit as the average Water Content of the three determinations and report the Plastic Limit to the nearest whole number.

Calculations

For each determination of Plastic Limit, determine the Water Content using the procedures described in Chapter 1. Report the Plastic Limit as the mean value of the three determinations, to the nearest integer.

Example

The results in Table 12.3 were obtained from a sample of Buckshot Clay using the rolling thread method for determining Plastic Limit. The Plastic Limit is reported as: P.L. = 30.

Table 12.3 Sample of Buckshot Clay

Trial no.	Tare no.	Wet mass + tare	Dry mass + tare	Tare mass	W (%)
1	A	8.46	7.05	2.24	29.3
2	B	5.12	4.24	1.30	29.9
3	12	6.56	5.37	1.35	29.6

Discussion

Considerable experience is needed in order to obtain consistent and reliable results for Plastic Limit using the thread test. In addition to being variable, another difficulty with the rolled thread method is the difficulty at even defining the Plastic Limit for soils of low clay content and other troublesome soils, leading to a definition of NP (nonplastic).

Other suggestions have been made to improve the thread method of determining the Plastic Limit using a plastic rolling device (e.g., Bobrowski & Griekspoor 1992; Ishaque et al. 2010; Kayabali 2012; Moreno-Maroto and Alonso-Azcarate 2015). In fact, such a device is allowed in ASTM D4318. However, these methods appear to be less common in practice than using the hand-rolled thread method.

PLASTIC LIMIT BY FALL CONE

Introduction

Because many engineers feel that determining the Plastic Limit by the thread test is very subjective on the part of the operator, it has been suggested that the Plastic Limit be redefined as the Water Content at which there is a 100-fold increase in undrained shear strength. As previously discussed, the undrained shear strength of the soil can be determined using the Fall Cone. If the Fall Cone gives a value of undrained shear strength of 1.6 kPa at the Liquid Limit, then the Plastic Limit is defined as the Water Content giving an undrained shear strength of 160 kPa. In order to actually determine this value with a Fall Cone, the mass of the cone needs to be increased and the penetration depth reduced since neither a 60 gm nor an 80 gm cone will penetrate far enough into the soil to obtain a high enough value of strength.

Standard

There currently is no ASTM standard for determining the Plastic Limit using the Fall Cone, although the process of developing a test standard has been initiated.

Equipment

Fall Cone device
Cone with different angles and mass (up to 400 gm)
Ceramic or plastic mixing bowls
No. 40 sieve
Ceramic mortar and rubber-tipped pestle
Electronic scale with a minimum capacity of 250 gm and a resolution of 0.01 gm
Convection oven capable of maintaining 110°±5°C
Distilled water
Plastic wrap
Spatula
Ceramic or aluminum moisture tares
Stop watch

Procedure

1. Follow the test procedure previously described for determining the Liquid Limit using the Fall Cone. Use the same soil sample prepared for the Liquid Limit test.
2. Allow the soil to dry below the Liquid Limit and continue obtaining penetration measurements using cones of appropriate mass on samples placed in the cup.
3. For each measurement of penetration, obtain a sample of about 10–15 gm from the middle of the cup and determine the oven dry Water Content.
4. Prepare a semi-log plot of Water Content vs. undrained shear strength.

Calculations

1. Calculate the undrained shear strength from:

$$s_u = 9.8\big[(M_c)(K)\big]\big/d^2 \tag{12.2}$$

where:
s_u = undrained shear strength (kPa)
M_c = mass of the cone (gm)
K = a constant depending on cone apex angle ($K = 0.8$ for 30° cone;
    $K = 0.27$ for 60° cone)
D = depth of penetration (mm)

Example

Fall Cone tests were performed on remolded Buckshot Clay. Cones with mass of 10 gm, 60 gm, 100 gm and 400 gm were used. The resulting relationship between Water Content and undrained shear strength is shown in Figure 12.16. From these data, the Water Content corresponding to $s_u = 1.6$ kPa = L.L. = 56% and the Water Content corresponding to $s_u = 160$ kPa = P.L. = 26%.

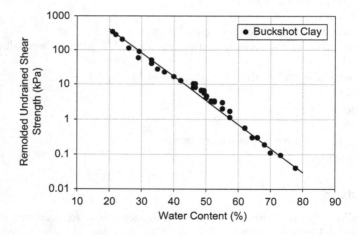

Figure 12.16 Relationship between Water Content and remolded undrained shear strength.

Discussion

The use of a Fall Cone to define the Plastic Limit as the Water Content corresponding to an undrained shear strength 100 times the Liquid Limit eliminates the problem encountered in very low plasticity clays and silts in which the thread method doesn't work. It appears that Youseff et al. (1965) were the first to suggest that the Plastic Limit may also be obtained from the Fall Cone much in the same way as the Liquid Limit is. The results shown in Figure 12.16 also show that when plotted on a semi-log plot, the data show a linear relationship. This is a characteristic behavior of nearly all fine-grained soils and helps explain why the data from the Casagrande drop cup also show a linear relationship when plotted on a semi-log plot. Dropping the Casagrande cup onto the hard base is a form of dynamic shear strength test. Additional discussion of the use of the Fall Cone to determine Plastic Limit has been presented by Campbell 1976; Harison 1988; Stone and Phan 1995; Muntohar and Hashim 2005; and Haigh et al. 2013.

ONE-POINT LIQUID LIMIT TESTS

A way to reduce the testing time needed to determine the Liquid Limit is to use a procedure referred to as the *One-Point Liquid Limit test*. Since the late 1940s, a number of studies have been performed to expedite the determination of the Liquid Limit by using a One-Point test method. ASTM D4318 (Method B) allows determination of the Liquid Limit based largely on work performed by the U.S. Waterways experiment station and Olmstead (1970). The soil–water mixture is prepared to give between 20 and 30 drops of the Casagrande drop cup to close the groove. The Liquid Limit may then be determined from:

$$LL_N = W(N/25)^{0.121} \qquad (12.5)$$

where:
    W = Water Content corresponding to N
    N = number of drops of the cup

or

$$LL = k \times N$$

where;
    k = a factor given in Table 12.4.

In the U.K., the British standard test procedure uses a similar equation based on the work by Norman (1959):

$$LL = W_N (N/25)^{0.092} \qquad (12.6)$$

Nagaraj and Jayadeva (1981) used previous work by Clayton and Jukes (1978); Karlsson (1961) and Sherwood and Ryley (1970) to suggest the

Table 12.4 Factors for Obtaining Liquid Limit from One-Point Test

N	k
20	0.973
21	0.979
22	0.985
23	0.990
24	0.995
25	1.000
26	1.005
27	1.009
28	1.014
29	1.019
30	1.022

following equations for determining the Liquid Limit from the Fall Cone using a One-Point method:

$$L.L. = W_P/(0.77 \log P) \tag{12.7}$$

$$L.L. = W_P/(0.65 + 0.0175P) \tag{12.8}$$

where:
W_P = Water Content corresponding to penetration, P
P = cone penetration (mm)

One-Point Liquid Limit tests are useful when there is only a small quantity of soil available or when time is short and results are needed quickly. However, when possible, the full Liquid Limit test should be performed.

DISCUSSION OF USING ATTERBERG LIMITS

There are some important indices that can be defined using the Atterberg Limits that help describe the nature of fine-grained soils. these include Plasticity Index, Shrinkage Index, Liquidity Index and Consistency Index:

$$\text{Plasticity Index}(P.I.) = L.L. - P.L. \tag{12.9}$$

$$\text{Shrinkage Index}(S.I.) = L.L. - S.L. \tag{12.10}$$

$$\text{Liquidity Index}(L.I.) = [((W - P.L.)]/P.I. \tag{12.11}$$

$$\text{Consistency Index}(I.C.) = [(L.L. - W)]/P.I. \tag{12.12}$$

Using Equations 12.11 and 12.12, we may develop an identity between the Liquidity Index and the Consistency Index as:

$$L.I. = 1 - I.C. \qquad (12.13)$$

From Equation 12.13, if the Water Content of a soil is equal to the Liquid Limit, the Liquidity Index is equal to 1.0, indicating a very soft soil. If the natural Water Content of a soil is equal to the Plastic Limit, the Liquidity Index is equal to 0.0, indicating a very stiff soil. Negative values of Liquidity Index are possible if the natural Water Content is below the Plastic Limit. Values of Liquidity Index between 0 and 1 are encountered in many fine-grained soils, but values greater than 1.0 suggest very soft and often highly sensitive soils. Negative values of Liquidity Index sometimes occur in highly expansive or swelling soils or other very stiff clays above the water table. Table 12.5 gives some results of Atterberg Limits determined for a variety of fine-grained soils.

Aterberg Limits and Unified Soil Classification

Fine-grained soils are traditionally thought of as "silts" or "clays", the names primarily related to specific grain-size designations. Clay is defined as particles with a diameter less than 0.002 mm. Silt falls in between sand and clay with particles between 0.074 mm and 0.002 mm. The terms *silty-clay* or *clayey-silt* imply soils that contain a range of particles within the silt and clay size but with behavior that tends more toward clay or silt.

The Liquid and Plastic Limits are important indices and are a primary part of the Unified Soil Classification System (USCS) traditionally used for classification of fine-grained soils. They are used in conjunction with the

Table 12.5 Typical Values of Atterberg Limits for a Variety of Fine-Grained Soils

Soil	L.L.	P.L.	S.L.	P.I.	S.I.
Peerless Clay (Kaolinite)	60	32	30	28	30
Theile Kaolin	65	38	35	27	30
Texas Ca Montmorillonite	142	44	15	98	127
Wyoming Bentonite	519	35	14	484	505
Boston Blue Clay	39	20	20	19	19
Leda Clay	48	27	21	21	27
Houston Clay	61	28	14	33	47
Salt Lake City Clay	25	17	22	8	3
Omaha Loess	44	17	15	25	27
London Clay	60	27	14	33	46
Osaka Japan Clay	105	40	37	65	68
Gault Clay	74	19	15	55	59

Casagrande Plasticity Chart to separate different fine-grained soils according to expected behavior using designated boundaries. The Casagrande Chart is a plot of Plasticity Index versus Liquid Limit as shown in Figure 12.17. The distinction between low plasticity soils and high plasticity soils is Liquid Limit = 50. The "A-Line" generally represents the separation between "clays" and "silts", with clays plotting above the A-Line and silts below. The A-Line has the equation P.I. = 073 (L.L − 20); this gives a P.I. = 0 at LL = 20. This means that the USCS identifies 6 basic soil types:

OL – organic soil with low plasticity
OH – organic soil with high plasticity
ML – low plasticity silts
MH – high plasticity silts
CL – low plasticity clay
CH – high plasticity clay

In addition to these six soil types, there is a provision for some soils that are borderline and have a dual designation, e.g., OL-ML; CL-OL; etc. Note, however, there is no provision in the USCS system for "medium plasticity" soils. Should it be expected that a soil magically changes from *low* plasticity to *high* plasticity as the Liquid Limit crosses the value of 50?

Partly because of the complexity of fine-grained soils and the simplicity of the Casagrande Plasticity Chart, several suggestions have been made to improve the chart and provide additional soil classification categories. For example, Polidori (2003) suggested the addition of the "C-Line", which is

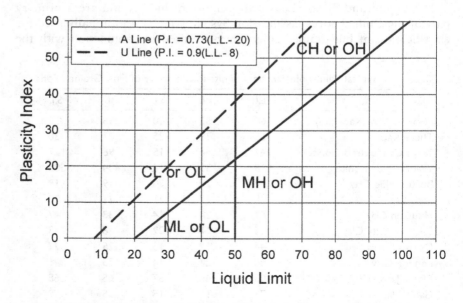

Figure 12.17 Casagrande USCS Plasticity Chart.

parallel to but lies below the Casagrande A-Line and identifies soils dominated by non-platy minerals, such as halloysite and allophane.

The British Soil Classification System (BSCS) recognizes ten different classes of soil, still separated by Liquid Limit, allowing more detailed distinction between soils, as shown in Table 12.6. This approach seems more logical and allows for more detailed separation of soils.

The European Soil Classification System (ESCS), which is used in some parts of the world, uses the basic Casagrande Plasticity Chart, but recognizes different plasticity characteristics for inorganic silt and clay based primarily on the position relative to the A-Line on the plasticity chart. Table 12.7 gives soil classification for six different fine-grained soils according to the ESCS.

Several comparisons and discussions have been presented recently comparing the different soil classification systems for fine-grained soils (Kovacevic & Juric-Kaconic 2004; Kovacevic et al. 2016; Okkels 2019; Afgolagboye et al. 2021). Some soils, especially lateritic and tropical soils containing allophane or halloysite, fall well below the A-Line. As noted, some other plasticity charts and soil classification systems have been suggested, most recently

Table 12.6 Soil Classification by the British Soil Classification System (BSCS)

Designation	Description	Range of L.L.
ML	Low plasticity silt	< 35
MI	Intermediate plasticity silt	35–50
MH	High plasticity silt	50–70
MVH	Very high plasticity silt	70–90
MEH	Extremely high plasticity silt	90+
CL	Low plasticity clay	< 35
CI	Intermediate plasticity clay	35–50
CH	High plasticity clay	50–70
CV	Very high plasticity clay	70–90
CE	Extremely high plasticity clay	90+

Table 12.7 Soil Classification According to the ESCS

Liquid Limit	Position Relative to the A-Line	Soil Classification	Symbol
< 35	At or Above the A-Line	Low Plasticity Clay	CIL
	Below the A-Line	Low Plasticity Silt	SiL
35-50	At or Above the A-Line	Medium Plasticity Clay	CII
	Below the A-Line	Medium Plasticity Silt	SiI
>50	At or Above the A-Line	High Plasticity Clay	CIH
	Below the A-Line	High Plasticity Silt	SiH

by Polidori (2003), Vardanega et al. (2022) and Moreno-Maroto et al. (2021). Some of these systems make use of Fall Cone results or flow index.

Activity

Another important use of the Plasticity Index is to define the Activity of the soil which combines the Atterberg Liquid and Plastic Limits and the grain-size characteristics (clay content). Skempton (1953) defined Activity as:

$$A = P.I./\% < 0.002\,mm \qquad (12.14)$$

Equation 12.14 says that Activity is simply the slope of a plot of Plasticity Index vs. clay fraction. The Activity provides a general indication of the clay mineralogy of fine-grained soils and can be a useful means of estimating the presence of swelling expansive clays at a site (discussed further in Chapter 16). The range of Activity values for most soils is from about 0.4 for Kaolinite to about 8.0 for montmorillonite. Most natural soils have values ranging from about 0.4 to 2.0. Typical values of Activity for a variety of different soils are given in Table 12.8.

Plasticity Ratio

The Atterberg Plastic and Liquid Limits may also be used to provide an initial indication of the clay mineralogy of fine-grained soils. The ratio of P.L./L.L. is called the Plasticity Ratio. The ratio tends to have very low values (0.1–0.2) for soils with montmorillonitic clay minerals and values closer to 1.0 for soils dominated by kaolinite. Figure 12.18 shows some results of tests on two-clay mineral mixes.

Table 12.8 Typical Values of Activity for a Variety of Soils

Soil	P.I.	Clay content (% < 0.002 mm)	Activity (P.I./CF)
Peerless Clay (Kaolinite)	28	76.0	0.44
Theile Kaolin	27	36.2	0.75
Texas Ca Montmorillonite	98	73.3	1.34
Wyoming Bentonite	484	60.4	8.01
Boston Blue Clay	19	53.0	0.36
Leda Clay	21	69.0	0.30
Houston Clay	33	71.1	0.46
Salt Lake City Clay	28	25.1	1.12
Omaha Loess	25	12.4	2.02
London Clay	33	60.4	0.55
Osaka Japan Clay	65	36.6	1.78
Gault Clay	55	58.3	0.94

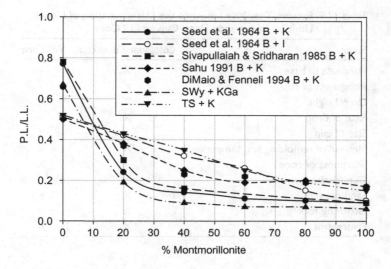

Figure 12.18 Plasticity Ratio as a function of clay composition.

Discussion

It should be kept in mind that the Aterberg Limits are simple index values based on simple tests using remolded soil and they have some inherent variability. In some cases, their use has been applied to achieve a false sense of both precision and accuracy for predicting other soil behavioral characteristics. They help to indirectly identify basic soil behavior but should not be considered as a substitute for direct tests to determine specific soil behavior. They can often be used with great success as a screening test to suggest that more detailed laboratory tests may be needed for a specific project.

Several investigations have shown that pretreating some soils by either air-drying or even oven-drying prior to performing the Atterberg Limits tests can affect the test results (Birrell 1952; Muller-Vonmoos 1965; Moh & Mazhar 1969; Wesley 1973; Sangrey et al. 1976; Prusza et al. 1983; Armstrong & Petry 1986; Rao et al. 1989; Pandian et al. 1991; Hight et al. 1992; Basma et al. 1994; Sridharah 1999; Low & Phoon 2003; Huvaj & Uyeturk 2018). In general, air drying and oven drying tend to reduce the Liquid Limit and Plastic Limit by as much as 50%; depending on mineralogy and Organic Content, soils with halloysite and allophane (lateritic and volcanic soils) and marine deposits show the largest change in liquid and Plastic Limit from natural Water Content to air drying.

These changes can shift the soil from one class to another and can change other characteristics, such as Activity. When possible, Atterberg Limits should be performed on samples at natural Water Content. There may be a number of factors that influence the determination of Liquid Limit using

Table 12.9 Factors That May Influence the Determination of Liquid
Limit by Casagrande Drop Cup or Fall Cone

Factor	Casagrande drop cup	Fall Cone
Hardness of base	X	-
Roughness of cup	X	-
Drop height	X	X
Rate of drop	X	-
Size of cup	-	X
Method of remolding and tempering	X	X
Sharpness of cone	-	X
Cup friction	X	-
Sand content	X	X
Grooving tool	X	
Pretreatment	X	X

either the Casagrande drop cup or the Fall Cone. Table 12.9 lists a number of these factors.

The concepts of soil plasticity following the work of Atterberg and Casagrande have recently undergone some reconsideration and perhaps overdue updating. As the understanding of the soil factors that influence Liquid and Plastic Limit, such as undrained shear strength, soil suction, SSA, CEC, amorphous content, Carbonate Content, Organic Content, etc., become better understood, the use of Atterberg Limits may expand. Nonetheless, they are useful indices that should be routinely performed as part of a site investigation to help understand soil behavior.

REFERENCES

Afgolagboye, L. O., Talabi, A.O. and Owoyemi, O.O., 2021. The Use of Polodori's Plasticity and Activity Charts in Classifying Some Residual Lateritic Soils in Nigeria. *Heliyon*, Vol, 7, No 8

Armstrong, J.C., and Petry, T.M., 1986. Significance of Specimen Preparation Upon Soil Plasticity. *ASTM Geotechnical Testing Journal*, Vol. 9, No. 3, pp. 147–153.

Basma, A.A., Al-Homoud, A.S., and Al-Tabari, E.Y., 1994. Effects of Methods of Drying on the Engineering Behavior of Clays. *Applied Clay Science*, Vol. 9, No. 3, pp. 151–164.

Birrell, K.S., 1952. Some Physical Properties of New Zealand Volcanic Ash Soils. *Proceedings of the 1st Australia-New Zealand Conference on Soil Mechanics and Foundation Engineering*, pp. 30–34.

Bjerrum, L., and Flodin, N., 1960. The Development of Soil Mechanics in Sweden, 1900–1925. *Geotechnique*, Vol. 10, No. 1, pp. 1–18.

Bobroiwski, L.J., and Griekspoor, D.M., 1992. Determination of the Plastic Limit of a Soil by Means of a Rolling Device. *ASTM Geotechnical Testing Journal*, Vol. 15, No. 3, pp. 284–287.

Bretting, A.E., 1936. Soil Studies for the Storstrom Bridge, Denmark. *Proceedings of the 1st International Conference on Soil Mechanics and Foundation Engineering*, Vol. 1, pp. 314–321.

Casagrande, A., 1932. Research on the Atterberg Limits of Soils. *Public Roads*, Vol. 13, No. 8, pp. 121–136.

Clayton, C.R., and Jukes, A.W., 1978. A One Point Cone Penetrometer Liquid Limit Test? *Geotechnique*, Vol. 28, No. 4, pp. 469–472.

Farrell, E., 1997. ETC 5 Fall-Cone Study. *Ground Engineering*, Vol. 30, No, 1, pp. 33–35.

Haigh, S.K., Vardenaga, P.J., and Bolton, M.D., 2013. The Plastic Limit of Clays. *Geotechnique*, Vol. 63, No. 6, pp. 435–440.

Hansbo, S., 1957. A New Approach to the Determination of the Shear Strength of Clays by the Fall-Cone Test. *Proceedings of the Royal Swedish Geotechnical Institute*, Vol. 14, pp. 7–48.

Harison, J.A., 1988. Using the BS Cone Penetrometer for the Determination of the Plastic Limit of Soils. *Geotechnique*, Vol. 38, pp. 433–438.

Hight, D.W., Bond, A.J., and Legge, J.D., 1992. Characterization of the Bothkennar Clay: An Overview. *Geotechnique*, Vol. 42, No. 2, pp. 303–347.

Huvaj, N., and Uyeturk, E., 2018. Effects of Drying on Atterberg Limits of Pyroclastic Soils of North Turkey. *Applied Clay Science*, Vol. 162, pp. 46–56.

Ishaque, F., Hoque, M.N., and Rashid, M.A., 2010. Determination of Plastic Limit of Some Selected Soils Using Rolling Device. *Progressive Agriculture*, Vol. 21, Nos. 1–2, pp. 187–194.

Karlsson, R., 1961. Suggested Improvements in the Liquid Limit Test with Reference to Flow Properties of Remolded Clays. *Proceedings of the 5th International Conference on Soil Mechanics and Foundation Engineering*, Vol. 1, pp. 171–184.

Kayabali, K., 2012. An Alternative Testing Tool for Plastic Limit. *Electronic Journal of Geotechnical Engineering*, Vol. 17, pp. 2107–2114.

Kovacevic, M. S., and Junic-Kacunic, D., 2014. European Soil Classification System for Engineering Purposes. *GRADEVINAR*, Vol. 66, pp. 801–810.

Kovacevic, M. S., Junic-Kacunic, D., and Libric, L., 2016. Comparison of Unified and European Soil Classification Systems. *Geotechnical and Geophysical Site Characterization*, Vol. 5, pp. 285–289.

Leroueil, S., and Le Bihan, J.P., 1996. Liquid Limits and Fall Cones. *Canadian Geotechnical Journal*, Vol. 33, pp. 793–798.

Low, H.E., and Phoon, K.K., 2003. Effect of Drying on the Liquid Limit of Singapore Marine Clay. *Proceedings of the 12th Asian Regional Conference on Soil Mechanics and Geotechnical Engineering*, Vol. 1, pp. 51–54.

Moh, Z.C., and Mazhar, M.F., 1969. Effects of Method of Preparation on Index properties of Lateritic Soils. *Proceedings of the 7th International Conference on Soil Mechanics and Foundation Engineering, Proceedings of Specialty Session on Engineering Properties of Lateritic Soils*, pp. 23–35.

Moreno-Maroto, J.M., and Alonzo-Azcarate, J., 2015. An Accurate, Quick and Simple Method to Determine Plastic Limit and Consistency Changes in all

Types of Clay and Soil: The Thread Bending Test. *Applied Clay Science*, Vol. 114, pp. 497–508.

Moreno-Maroto, J.M., Alonzo-Azcarate, J., and O'Kelly, B.C., 2021. Review and Critical Examination of Fine-Grained Soil Classification Systems Based on Plasticity. *Applied Clay Science*, Vol. 200, No. 1, pp. 1–13.

Muller-Vonmoos, M., 1965. Determination of Organic Matter for the Classification of Soil Samples. *Proceedings of the 6th International Conference on Soil Mechanics and Foundation Engineering*, Vol. 1, pp. 77–79.

Muntohar, A.S., and Hashim, R., 2005. Determination of Plastic Limit of Soils Using Cone Penetrometer: Re-Appraisal. *Jurnal Teknik Sipil*, Vol. 11, No. 2, p. 11.

Nagaraj, T.S., and Jayadeva, M.S., 1981. Re-Examination of One-Point Methods of Liquid Limit Determination. *Geotechnique*, Vol. 31, No. 4, pp. 413–425.

Norman, L.E.J., 1959. The One-Point Method of Determining the Value of the Liquid Limit of a Soil. *Geotechnique*, Vol. 9, No. 1, pp. 1–8.

Okkels, N., 2019. Modern Guidelines for Classification of Fine Soils. *Proceedings of the 17th European Conference on Soil Mechanics and Geotechnical Engineering*, 9 pp.

Olmstead, F.R., 1970. Suggested Methods of Test for Securing the Liquid Limit of Soils Using One-Point Data. *ASTM STP*, Vol. 479, pp. 94–96.

Pandian, N.S., Nagaraj, T.S., and Sivakumar Babu, S.L., 1991. Effects of Drying on the Engineering Behavior of Cochin Marine Clays. *Geotechnique*, Vol. 41, No. 1, pp. 143–147.

Polidori, E., 2003. Proposal for New Plasticity Chart. *Geotechnique*, Vol. 53, No. 4, pp. 397–406.

Pruza, Z., Kleiner, D.E., and Sundaram, A.V., 1983. Characteristics of Guri Soils. *Geologic Environment and Soil Properties, ASCE*, pp. 183–199.

Rao, S.M., Sridharah, A., and Chandrakaran, S., 1989. Influence of Drying on the Liquid Limit Behavior of a Marine Clay. *Geotechnique*, Vol. 39, No. 4, pp. 715–175.

Sangrey, D.A., Noonan, D.K., and Webb, G.S., 1976. Variation in Atterberg Limits of Soils Due to Hydration History and Specimen Preparation. *ASTM STP*, Vol. 599, pp. 158–165.

Sherwood, P.T., and Ryley, M.D., 1970. An Investigation of a Cone Penetrometer Method for Determination of the Liquid Limit. *Geotechnique*, Vol. 20, No. 2, pp. 203–208.

Shridharan, A., 1999. Engineering Behavior of Marine Clays. *Proceedings of the International Conference on Offshore and Nearshore Geotechnical Engineering*, pp. 49–63.

Skempton, A.W., 1953. The Colloidal Activity of Clays. *Proceedings of the 3rd International Conference on Soil Mechanics and Foundation Engineering*, Vol. 1, pp. 57–61.

Skempton, A.W., 1985. A History of Soil Properties, 1717–1927. *Proceedings of the 11th International Conference on Soil Mechanics and Foundation Engineering, Golden Jubilee Volume*, pp. 95–125.

Stone, K.J.L. and Phan, K.D., 1995. Cone Penetration Tests Near the Plastic Limit. *Geotechnique*, Vol. 45, No. 1, pp. 155–158.

Vardanega, P.J., Haigh, S.K., and O'Kelly, B.C., 2022. Use of Fall-Cone Flow Index for Soil Classification: A New Plasticity Chart. *Geotechnique*, Vol. 72 No. 7 pp. 610–617

Wesley, L.D., 1973. Some Basic Engineering Properties of Halloysite and Allophane Clays in Java, Indonesia. *Geotechnique*, Vol. 23, No. 4, pp. 471–494.

Youssef, M.S., El Ramli, A.H., and El Demery, M., 1965. Relationships Between Shear Strength, Consolidation, Liquid Limit and Plastic Limit for Remoulded Clays. Proceedings of the 6th International Conference on Soil Mechanics and Foundation Engineering, Vol. 1, pp. 126–129.

ADDITIONAL READING

Belviso, R., Ciampoli, S., Cotecchia, V., and Federico, A., 1985. Use of the Cone Penetrometer to Determine Consistency Limits. *Ground Engineering*, Vol. 18, No. 5, pp. 21–22.

Brown, P.J., and Huxley, M.A., 1996. The Cone Factor for a 30° Cone. *Ground Engineering*, Vol. 29, No. 10, pp. 34–36.

Budhu, M., 1985. The Effect of Clay Content on Liquid Limit from Fall Cone and British Cup Device. *Geotechnical Testing Journal, ASTM*, Vol. 8, No. 2, pp. 91–95.

Campbell, D.J., 1976. Plastic Limit Determination Using a Drop Cone Penetrometer. *Journal of Soil Science*, Vol. 27, pp. 295–300.

Casagrande, A., 1948. Classification and Identification of Soils. *Transactions of the ASCE*, Vol. 113, pp. 901–930.

Claveau-Mallet, D., Duhaime, F., and Chapuis, R.P., 2011. Characterization of Champlain Saline Clay from Lachenaie Using the Swedish Fall Cone. *Proceedings of GeoCalgary*, pp. 134–138.

Claveau-Mallet, D., Duhaime, F., and Chapuis, R.P., 2012. Practical Considerations When Using the Swedish Fall Cone. *ASTM Geotechnical Testing Journal*, Vol. 35, No. 4, pp. 1–11.

Di Matteo, L., 2012. Liquid Limit of Low- to Medium-Plasticity Soils: Comparison Between Casagrande Cup and Cone Penetrometer Test. *Bulletin of Engineering Geology and Environment*, Vol. 71, pp. 79–85.

Di Maio, C., and Fenelli, G.B., 1994. Residual Strength of Kaolin and Bentonite: The Influence of their Constituent Pore Fluid. *Geotechnique*, Vol. 44, No. 2, pp. 217–226.

El-Shinawi, A., 2017. A Comparison of Liquid Limit Values for Fine Soils: A Case Study at the North Cairo-Suez District, Egypt. *Journal of the Geological Society of India*, Vol. 89, pp. 339–343.

Farrell, E., Schuppener, B., and Wassing, B., 1997. ETC Fall-Cone Study. *Ground Engineering*, Vol. 30, No. 1, pp. 33–36.

Federico, A., 1983. Relationships $(c_u - w)$ and $(c_u - \delta)$ for Remoulded Clayey Soils at High Water Content. *Italian Geotechnical Engineering Journal*, Vol. 17, No. 1, pp. 38–41.

Feng, T.W., 2000. Fall-Cone Penetration and Water Content Relationships of Clays. *Geotechnique*, Vol. 50, No. 2, pp. 181–187.

Golawska, K, Lechowicz, Z., Matusiewicz, W., and Sulewska, M.J., 2020. Determination of the Atterberg Limits of Eemian Gytta on samples with Different Composition. *Studia Geotechnica et Mechanica*, Vol. 42, No. 2, pp. 168–178.

Gronbech, G.L., Nielsen, B.N., and Ibsen, L.B., 2011. Comparison of Liquid Limit of Highly Plastic Clay by Means of Casagrande and Fall Cone Apparatus. *Proceedings of the 2011 Canadian Geotechnical Conference.*

Ibrahim, H.H., and Noori, K.M., 2019. Determining Casagrande Liquid Limit Values from Cone Penetration Test Data. *Journal of Pure and Applied Sciences*, Vol. 31, No. S3, pp. 114–120.

Lu, T., and Bryant, W.R., 1997. Comparison of Vane Shear and Fall Cone Strengths of Soft Marine Clay. *Marine Georesources and Geotechnology*, Vol. 15, pp. 67–82.

Prakash, K., and Sridharan, A., 2006. Critical Appraisal of the Cone Penetration Method of Determining Soil Plasticity. *Canadian Geotechnical Journal*, Vol. 43, pp. 884–888.

Sahu, B.K., 1991. Atterberg Limits of Soil Mixtures. *Proceedings of the 9th Asian Regional Conference on Soil Mechanics and Foundation Engineering*, Vol. 1, pp. 163–166.

Seed, H.B., Woodward, R.J., and Lundgren, R., 1964. Clay Mineralogical Aspects of the Atterberg Limits Tests. *Journal of the Soil Mechanics and Foundations Division, ASCE*, Vol. 90, No. 4, pp. 107–131.

Shridharan, A., El-Shafei, A., and Miura, N., 2002. Mechanisms Controlling the Undrained Shear Strength Behavior of Remolded Ariake Marine Clays. *Marine Georesources and Geotechnology*, Vol. 20, pp. 21–50.

Sivapullaiah, P.V., and Sridharan, A., 1985. Liquid Limit of Soil Mixtures. *ASTM Geotechnical Testing Journal*, Vol. 8, No. 3, pp. 111–116.

Wasti, Y., and Bezirci, M.H., 1986. Determination of the Consistency Limits of Soils by the Fall Cone Test. *Canadian Geotechnical Journal*, Vol. 23, pp. 241–246.

Wan, Y., Kwong, J., Brandes, H.G., and Jones, R.C., 2002. Influence of Amorphous Clay-Size Materials on Soil Plasticity and Shrink-Swell Behavior. *Journal of Geotechnical and Geoenvironmental Engineering, ASCE*, Vol. 128, No. 12, pp. 1026–1031.

Wasti, Y., 1987. Liquid and Plastic Limits as Determined from the Fall Cone and the Casagrande Methods. *Geotechnical Testing Journal, ASTM*, Vol. 10, No. 1, pp. 26–30.

Chapter 13

Shrinkage Limit

BACKGROUND

Most fine-grained soils shrink when water is removed by drying. Shrinkage can occur when the soil is exposed to the air and dries naturally or it can shrink if water is removed from the soil by vegetation, such as trees around houses. The shrinkage produces a decrease in volume of the soil. These soils are often called *shrink–swell* soils and they can create severe problems for lightly loaded structures such as houses. Engineers use a number of relatively simple and quick tests to estimate whether a particular soil is prone to shrink–swell behavior and to what degree or how severe the swelling or shrinking might be. One simple test that can be used to describe the soil's shrink characteristics is the Shrinkage Limit test, which was one of the plasticity limits defined by Atterberg.

Figure 13.1 illustrates the change in volume associated with the change of Water Content at the various Atterberg Limits. Conceptually, the volume decreases linearly from the Liquid Limit (L.L.) through to Plastic Limit (P.L.) to the Shrinkage Limit (S.L.). The soil is saturated at Water Content above the Liquid Limit and remains saturated as it loses water and shrinks down to the Plastic Limit. Further drying past the Plastic Limit causes a small amount of shrinkage, but then at some point the shrinkage stops, the soil becomes unsaturated and the volume becomes constant, even though there is some water still in the soil.

The Shrinkage Limit is defined as the Water Content below which soil volume is constant. Any further reduction in Water Content will not produce a reduction in volume. As indicated in Figure 13.1, the *interpreted* Shrinkage Limit is usually defined by the intersection of two straight line segments of the drying curve. This definition of Shrinkage Limit is more-or-less a matter of tradition and convenience, but in reality, the soil can still experience a small amount of volumetric shrinkage below this value, as indicated by the drying curve, so that the *true* Shrinkage Limit actually

DOI: 10.1201/9781003263289-15

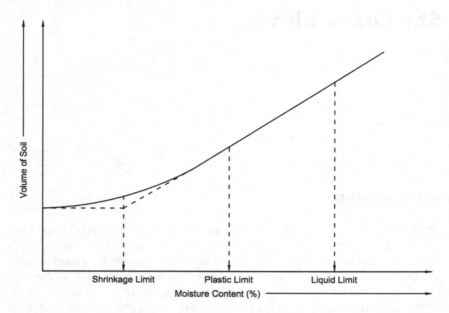

Figure 13.1 Ideal volume vs. Water Content of soil at Atterberg Limits.

occurs at the point of tangency between the curve at the lower straight-line segment.

Three methods that can be used to determine the Shrinkage Limit are described in this chapter. The first method described (Mercury Method) was an ASTM and ASSHTO standard for many years but was withdrawn by ASTM in 2008 because of potential hazards with handling mercury; however, it is still in use by some laboratories and was a very popular method for over 80 years.

SHRINKAGE LIMIT BY MERCURY METHOD

In this method, a soil–water paste is prepared and placed in a standardized dish called the Shrinkage Limit dish. The soil is allowed to air dry slowly and then the dried volume of the soil is determined by submerging the specimen in a mercury bath. The mass of the displaced mercury, which is a non-wetting fluid, is then measured and the volume is calculated from the mass using the Specific Gravity of mercury, giving the volume of the soil specimen (Figure 13.2).

Standard

ASTM D427 *Standard Test Method for Shrinkage Factors of Soils Limit by the Mercury Method*

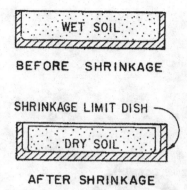

Figure 13.2 Volume change of soil in Shrinkage Limit dish.

Equipment

Porcelain or stainless steel Shrinkage Limit dish 1.75 in (44.4 mm) diameter by 0.5 in (12.7 mm high)

Glass cup 2.25 in (57.2 mm) diameter by 1.25 in (31.8 mm) high

Glass or Plexiglass plate with three steel prongs

Spatula

Mercury

Ceramic evaporating dish

Mercury

Convection oven capable of maintaining 110° ± 5°C

Electronic scale with a minimum capacity of 500 gm and a resolution of 0.1 gm

Straightedge

Ceramic or plastic mixing bowls

Distilled water (Figure 13.3)

Calibration of Shrinkage Limit dish

The volume of each Shrinkage Limit dish must be determined individually.

1. Place the dish in the bottom of a ceramic bowl.
2. Fill the dish with mercury until it overflows slightly.
3. Place a clean glass plate over the dish and press lightly to remove any excess mercury.
4. Pour the mercury from the Shrinkage Limit dish into a small bowl of known mass (M_{BOWL}) and determine the mass of the mercury and bowl ($M_{HG} + M_{BOWL}$).
5. Calculate the volume of the dish (which will also be equal to the volume of wet soil) using the mass of the mercury as:

Figure 13.3 Equipment needed for performing Mercury Method Shrinkage Limit test: shrinkage dish; glass cup; three prong plate.

$$V_{DISH} = V_{WET} = M_{HG}/\rho_M \tag{13.1}$$

where:

V_D = volume of the dish (cm³)

M_{HG} = mass of mercury (gm) = $(M_{HG} + M_{BOWL}) - M_{BOWL}$

ρ_M = unit density (Specific Gravity) of mercury = 13.6 gm/cm³

Procedure

1. Select a representative sample of soil for testing. If the soil is at natural Water Content, remove any particles larger than the No. 40 sieve. If the soil has been air dried, crush the soil using a ceramic mortar and rubber-tipped pestle and pass the soil through a No. 40 sieve.
2. Mix about 150 gm of soil with distilled water in a bowl until the consistency is soft enough to give about five drops of the Casagrande Liquid Limit cup to close the groove cut with the grooving tool. Make sure the soil–water mixture is mixed thoroughly. This will give a Water Content above the Liquid Limit.
3. Cover the bowl with plastic wrap and allow the mixture to temper for about 24 h. This will help the soil fully hydrate.

4. Coat the inside of a clean dry Shrinkage Limit dish with a very thin film of petroleum jelly or silicone grease and then determine the mass of the dish to the nearest 0.01 gm (M_{DISH}).

5. Place about 1/3 of the soil–water paste into the dish. Tap the dish on the countertop so that the soil flows to the edges of the dish and any air bubbles are removed.

6. Repeat Step 5 until the dish is full.

7. Use the metal straight edge or spatula to level off the surface of the soil. Clean the sides and bottom of the dish to remove any excess soil.

8. Determine the mass of the dish and wet soil ($M_{WET} + M_{DISH}$).

9. Set the dish aside and allow the soil to air dry until the color changes from dark to light. (Note: Air drying may take between 6 and 18 hours depending on the soil.)

10. After air drying, place the dish into a convection oven and dry for 24 h. at 110°C.

11. After oven drying, remove the dish from the oven and place it in a desiccator to cool. After cooling, determine the oven dry mass ($M_{DRY} + M_{DISH}$).

12. Carefully remove the soil pat from the dish.

13. Place a clean glass cup into a ceramic bowl or evaporating dish and carefully fill the glass cup with mercury.

14. Place the oven dry soil pat on top of the mercury and press down lightly with the three-prong plate until the plate is flush with the surface of the glass cup. This will displace an amount of mercury equal to the dry soil pat.

15. Remove the soil pat and glass cup from the bowl. Pour the displaced mercury into a small ceramic bowl of known mass and determine the mass to the nearest 0.01 gm.

16. The volume of the oven dried soil pat, V_{DRY}, can be calculated using Eq. 13.1.

17. Carefully dispose of the soil pat and the mercury in appropriate containers. (Note: The mercury can be can be reused for additional measurements.) (Figures 13.4 and 13.5)

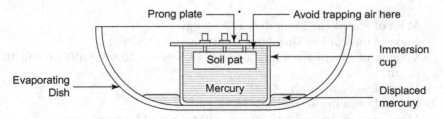

Figure 13.4 Placement of dry soil pat into glass cup filled with mercury.

Figure 13.5 Soil pat before and after drying and shrinking.

Calculations

Calculate the initial Water Content of the soil as:

$$Wi = \left[(M_{WET} - M_{DRY})/(M_{DRY} - M_{DISH}) \right] \times 100\% \qquad (13.2)$$

The initial volume of wet soil, V_{WET}, is equal to the volume of the Shrinkage Limit dish determined from the dish calibration.

Calculate the Shrinkage Limit, SL, as:

$$SL = Wi - \left[(V_{WET} - V_{DRY})\rho_W \right] / (M_{DRY}) \times 100\% \qquad (13.3)$$

where:

ρ_W = unit density of water = 1 gm/cm³

Report the Shrinkage Limit to the nearest whole number.

Example

The following test results were obtained from a kaolinitic clay:

Mass of Shrinkage Limit dish = 20.60 gm
Mass of mercury in Shrinkage Limit dish = 239.50 gm
Volume of Shrinkage Limit dish = (239.5 gm − 20.60 gm)/13.6 = 16.10 cm³
Mass of wet soil + dish = 46.48 gm
Mass of oven dry soil + dish = 35.99 gm
Mass of oven dry soil = 35.99 gm − 20.60 gm = 15.39 gm
Initial volume of soil = initial volume of dish = 16.10 cm³

Initial Water Content = [(46.48 gm − 35.99 gm)/(35.99 gm − 20.60 gm)]
    × 100% = 68.2%
Final volume of soil pat = 12.30 cm³

Using Equation 13.3:
    SL = 0.682 − [(16.10 cm³ − 12.30 cm³)1 gm/ cm³)]/15.39 gm = 0.247 = 25

Discussion

The Shrinkage Limit test is simple to perform, but technicians should wear
protective gloves and a face mask when handling the mercury. Errors in
the measurements occur from incorrectly determining the volumes of the
Shrinkage Limit dish or soil pat. Note: In some countries the use of Mercury
has been banned for determining the Shrinkage Limit and one of the more
direct methods should be used.

Another parameter that is often determined using the data obtained
in the Shrinkage Limit test is called the Shrinkage Ratio (SR), which is
an indicator of how much volume change may occur as changes in Water
Content above the Shrinkage Limit occur. The Shrinkage Ratio is defined
as the ratio of the volume change (expressed in terms of the dry volume)
to the corresponding change in Water Content above the Shrinkage Limit
(expressed in terms of the oven dry mass of the soil):

$$SR = (\Delta V/V_{DRY})/(\Delta W/Wi) \tag{13.4}$$

It can also be shown using soil phase relationships (Chapter 2) that:

$$SR = \rho_{DRY}/\rho_W \tag{13.5}$$

Equation 13.5 suggests that the higher the initial dry density of a given soil,
the more swelling will occur when the Water Content increases.

SHRINKAGE LIMIT BY WAX METHOD

Introduction

An alternative procedure for performing the Shrinkage Limit test uses wax
to encapsulate the soil specimen and eliminates the use of mercury. Similar
measurements are taken to obtain the initial and final Water Content and
initial and final volume of the soil.

Standard

ASTM D4943 *Standard Test Method for Shrinkage Factors by the Wax
Method*

Equipment

No. 40 sieve
Shrinkage Limit dish
Suspension apparatus
Silicone grease
Small spatula
Small ceramic bowl
Electronic scale with a minimum capacity of 500 gm and a resolution of 0.01 gm
Convection drying oven capable of maintaining $110° \pm 5°C$
Ceramic mortar and rubber-tipped pestle
Microcrystalline or other suitable wax (a 50/50 mixture of paraffin wax and petroleum jelly works well)
Distilled water
Fine sewing thread
Water bath (500 mL beaker)
Wax warmer (hot plate)
Pan from melting wax
Thermometer with a range of 0 to 25°C and a resolution of 0.5°C
Casagrande Liquid Limit drop cup
Glass plate (approximately 3.5 in × 3.5 in (80 mm × 80 mm))

Calibration of Shrinkage Limit dish

The volume of each Shrinkage Limit dish must be determined individually.

1. Apply a thin film of grease to the inside of a clean dry Shrinkage Limit dish and determine the mass to the nearest 0.01 gm (M_{DISH}).
2. Fill the Shrinkage Limit dish with water to overflowing.
3. Apply a thin film of grease to one face of a clean dry glass plate and determine the mass (M_{PLATE}). Place the plate over the dish and press lightly to remove any excess water.
4. Determine the mass of the greased dish, greased plate and water (M_{TOTAL}).
5. Calculate the mass of the water by subtracting the mass of the greased mold and greased glass plate from the total mass obtained in Step 4.
6. Calculate the volume of the dish (which will also be the volume of wet soil) using the mass of the water as:

$$V_{DISH} = V_{WET} = M_{WATER}/\rho_W \qquad (13.6)$$

where:

V_D = volume of the dish (cm³)
M_{WATER} = mass of water (gm)
ρ_W = unit density of water = 1 gm/cm³

(Note: An alternative procedure to determine the volume of the Shrinkage Limit dish is to measure the diameter and height and calculate the volume.)

Procedure

1. Select a representative sample of soil for testing. If the soil is at natural Water Content, remove any particles larger than the No. 40 sieve. If the soil has been air dried, crush the soil using a ceramic mortar and rubber-tipped pestle and pass the soil through a No. 40 sieve.

2. Mix about 150 gm of soil with distilled water in a bowl until the consistency is soft enough to give about five drops of the Casagrande drop cup to close the groove cut with the grooving tool. Make sure the soil–water mixture is mixed thoroughly. This will give a Water Content above the Liquid Limit.

3. Cover the bowl with plastic wrap and allow the mixture to temper for about 24 h. This will help the soil fully hydrate.

4. Coat the inside of the Shrinkage Limit dish with a very thin film of petroleum jelly or silicone grease and then determine the mass to the nearest 0.01 gm (M_{DISH}).

5. Place about 1/3 of the soil–water paste into the dish. Tap the dish on the countertop so that the soil flows to the edges of the dish and any air bubbles are removed.

6. Repeat Step 5 until the dish is full.

7. Use the metal straight edge or spatula to level off the surface of the soil. Clean the sides and bottom of the dish to remove any excess soil.

8. Determine the mass of the dish and wet soil ($M_{WET} + M_{DISH}$).

9. Set the dish aside and allow the soil to air dry until the color changes from dark to light. (Note: air drying may take between 6 and 18 hours depending on the soil.)

10. After air drying, place the dish into a convection oven and dry for 24 h at 110°C.

11. After oven drying, remove the dish from the oven and place in a desiccator to cool. After cooling, determine the oven dry mass ($M_{DRY} + M_{DISH}$).

12. Carefully remove the soil pat from the dish and determine the dry mass of soil as: $M_{DRY} = (M_{DRY} + M_{DISH}) - M_{DISH}$.

13. Securely tie sewing thread around the soil pat, leaving a thread of about 8 in (20 cm) long.

14. Immerse the soil pat into the molten wax using the sewing thread. Make sure the soil is completely covered. (Note: This should only take a few seconds.) If necessary, dip the soil two or three times, making sure that there are no air bubbles in the wax.

15. Allow the wax coating and soil to cool to room temperature.

16. Determine the mass of the wax coated soil pat to the nearest 0.01 gm ($M_{SOIL + WAX:AIR}$).

17. Suspend the wax coated soil pat from the thread and submerge the soil into a water bath placed on the scale. (Note: The water bath can be prepared using a 50 ml beaker filled with about 300 ml of water.). Determine the mass (M_{SOIL} + $_{WAX:WATER}$) to the nearest 0.01 gm.

Calculations

The initial Water Content of the soil is calculated as:

$$Wi = \left[(M_{WET} - M_{DRY})/(M_{DRY} - M_{DISH}) \right] \times 100\% \qquad (13.7)$$

The initial volume of wet soil, V_{WET}, is equal to the volume of the Shrinkage Limit dish determined from the dish calibration.

Calculate the volume of the dry soil pat as follows:

1. Calculate the mass of water displaced by the dry soil pat and wax as:

$$M_{WATER} = (M_{SOIL+WAX-AIR}) - (M_{SOIL+WAX-WATER})$$

2. Calculate the volume of the dry soil pat and wax as:

$$V_{SOIL+WAX} = M_{WATER} / \rho_W$$

ρ_W = unit density of water = 1 gm/cm^3

3. Calculate the mass of wax as:

$$M_{WAX} = (M_{DRY+WAX-AIR}) - M_{DRY}$$

4. Calculate the volume of wax as:

$$V_{WAX} = M_{WAX}/\rho_{wax}$$

ρ_{Wax} = unit density of wax

5. Calculate the volume of dry soil as:

$$V_{SOIL} = V_{SOIL+WAX} - V_{WAX}$$

Calculate the Shrinkage Limit, SL, as:

$$SL = Wi - \left[(V_{WET} - V_{DRY})\rho_W \right]/(M_{DRY}) \times 100\% \qquad (13.3)$$

where:

ρ_w = unit density of water = 1 gm/cm³

Report the Shrinkage Limit to the nearest whole number.

Discussion

In this method, the Specific Gravity of the wax, ρ_{wax}, must be known and usually can be obtained from the manufacturer. Prakash et al. (2009) reported that beeswax has ρ_{wax} = 0.9155 gm/cm³; Ito and Azam (2013) used a value of ρ_{wax} = 0.90 for microcrystalline wax. If using a mixture of 50/50 paraffin wax and petroleum jelly, the values for each component may be looked up online and then used to calculate the appropriate value. Comparisons of Shrinkage Limit values determined using the Mercury Method and the wax method were presented by Byers (1986) Prakash et al. (2009) and Kayabali (2013) and have shown that very similar results are obtained with both methods, probably within the error of repeatability.

SHRINKAGE LIMIT BY DIRECT MEASUREMENT METHOD

Introduction

This procedure for determining the Shrinkage Limit of soils may be performed along with either the Mercury Method or the wax method or it may be conducted by itself. Essentially, in this method, the wet soil pat is allowed to dry slowly, perhaps over several days, and periodically the size of the soil pat and its mass are determined. This procedure continues until there is no further reduction in dimensions of the pat, and then the soil and dish are placed in the oven to dry in order to obtain the dry mass of soil.

Standard

There is currently no ASTM standard for this test procedure. A similar procedure is covered in British Standard BS1377 – Part 2.

Equipment

Shrinkage Limit dish
Petroleum jelly or silicone grease
Small spatula
Small ceramic mixing bowl
Electronic scale with a minimum capacity of 250 gm and a resolution of 0.01 gm

Digital vernier caliper with a capacity of 150 mm (6 in) and a resolution of 0.1 mm (0.005 in)

Convection oven capable of maintaining $110° \pm 5°C$

Procedure

1. Select a representative sample of soil for testing. If the soil is at natural Water Content, remove any particles larger than the No. 40 sieve. If the soil has been air dried, crush the soil using a ceramic mortar and rubber-tipped pestle and pass the soil through a No. 40 sieve.
2. Mix about 150 gm of soil with distilled water in a bowl until the consistency is soft enough to give about five drops of the Casagrande drop cup to close the groove cut with the grooving tool. Make sure the soil–water mixture is mixed thoroughly. This will give a Water Content above the Liquid Limit.
3. Cover the bowl with plastic wrap and allow the mixture to temper for about 24 h. This will help the soil fully hydrate.
4. Coat the inside of the Shrinkage Limit dish with a very thin film of petroleum jelly or silicone grease and then determine the mass to the nearest 0.01 gm (M_{DISH}).
5. Place about 1/3 of the soil–water paste into the dish. Tap the dish on the countertop so that the soil flows to the edges of the dish and any air bubbles are removed.
6. Repeat Step 5 until the dish is full.
7. Use the metal straight edge or spatula to level off the surface of the soil. Clean the sides and bottom of the dish to remove any excess soil.
8. Determine the mass of the dish and wet soil ($M_{WET} + M_{DISH}$).
9. Set the dish aside and allow the soil to air dry slowly. (Note: Air drying may take between 6 and 18 hours depending on the soil.)
10. Periodically, at intervals of about 2–3 h, obtain the mass of the soil and dish. Also, obtain a measurement of the diameter of the soil pat using the digital calipers. Obtain three measurements of the diameter.
11. Continue this procedure until the soil pat stops shrinking and the diameter does not change. (Note: If the soil must be left overnight, cover the soil with plastic wrap so that the soil does not dry out too much overnight.)
12. After the soil has stopped shrinking, place the dish and soil in a convection oven for a minimum of 24 h at 110°C to oven dry the soil. After 24 h, determine the oven-dry mass ($M_{DRY} + M_{DISH}$) and obtain a final oven-dry measurement of the diameter of the soil pat.

Calculations

Using the individual measurements of mass, calculate the Water Content for each set of readings.

Calculate the average diameter of the soil pat for each set of readings.

Plot a graph of Water Content versus specimen diameter. Interpret the Shrinkage Limit as the point of intersection between the two straight line portions of the curve. Record the Water Content (as a whole number) at the intersection as the Shrinkage Limit.

Example

A sample of clay from Georgia was prepared and tested using the Shrinkage Limit dish to obtain a measure of Shrinkage Limit. The data are shown in Table 13.1. The initial Water Content of the soil was 58.5%.

A plot of the data is shown below in Figure 13.6. From the intersection of the straight-line segments, SL = 30.

Table 13.1 Sample of Clay from Georgia

Measurement no.	Diameter of soil pat (mm)	Water Content (%)
1	41.52	58.0
2	41.52	57.0
3	41.52	55.5
4	41.52	54.0
5	41.52	52.1
6	41.52	50.1
7	41.52	48.2
8	41.33	44.0
9	40.90	42.5
10	40.52	40.1
11	40.20	39.1
12	39.62	36.5
13	38.82	34.5
14	38.30	32.5
15	37.90	31.0
16	37.75	29.0
17	37.70	26.2
18	37.61	24.9
19	37.52	23.2
20	37.52	21.8
21	37.52	15.0
22	37.51	12.5
23	37.52	10.5
24	37.52	5.5
25	37.51	0.0

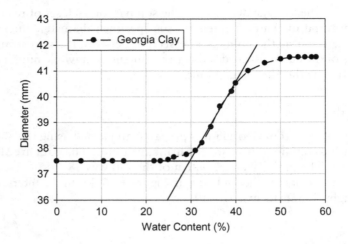

Figure 13.6 Plot of Water Content vs. diameter of drying soil pat in Shrinkage Limit dish.

Discussion

This method is considered a direct method for determining the Shrinkage Limit since it actually measures the point at which no further reduction in sample volume takes place. Results have been shown to provide good test results in comparison to conventional methods (Cerato & Lutenegger 2006). Also, the test does not use any hazardous materials, such as mercury.

Technically, even though the test procedures described in this chapter for determining Shrinkage Limit are based on using remolded soil prepared at high Water Content, it is possible to perform these tests using a small specimen of undisturbed soil (e.g., Hobbs et al. 2018). To do this, simply trim a soil specimen about the same size as the Shrinkage Limit dish and follow the procedures described in this chapter. The initial volume of the specimen will have to be determined by taking dimensions of the specimen using a digital caliper. After slowly air drying and then oven drying, the final volume can be determined using either the mercury, wax or direct measurement method. A discussion of this procedure is given in Chapter 15.

REFERENCES

Byers, J.G., 1986. Alternative Procedure for Determining Shrinkage Limit of Soil. U.U. Bureau of Reclamation Report No. REC-ERC-86-2.

Cerato, A.B., and Lutenegger, A.J., 2006. Shrinkage of Clays. *Proceedings of the 4th International Conference on Unsaturated Soils*, pp. 1097–1108.

Hobbs, P., Jones, L., Kirkham, M., Gunn, D., and Entwisle, D., 2018. Shrinkage Limit Test Results and Interpretation for Clay Soils. *Quarterly Journal of Engineering Geology*, Vol. 52, No. 2, pp. 220–229.

Ito, M., and Azam, S., 2013. Engineering Properties of a Vertisolic Expansive Soil Deposit. *Engineering Geology*, Vol. 152, pp. 10–16.

Kayabali, K., 2013. Evaluation of the Two Newly Proposed Methods for Shrinkage Limit. *Electronic Journal of Geotechnical Engineering*, Vol. 18, pp. 3047–3059.

Prakash, K., Sridharan, A., Anath Baba, J., and Thejas, H.K., 2009. Determination of Shrinkage Limit of Fine-Grained Soils by Wax Method. *Geotechnical Testing Journal*, Vol. 32, No. 1, pp. 86–89.

ADDITIONAL READING

Brasher, B.R., Franzmeier, D.P., Valassis, V., and Davidson, S.E., 1966. Use of Saran Resin to Coat Natural Soil Clods for Bulk-Density and Water Retention Measurements. *Soil Science*, Vol. 101, No. 2, pp. 108–109.

Izdebska-Mucha, D., and Wojcik, E., 2013. Testing Shrinkage Factors: Comparison and Correlation with Index Properties of Soils. *Bulletin of Engineering Geology and the Environment*, Vol. 72, pp. 15–24.

Sridharan, A., and Prakash, K., 1998. Mechanisms Controlling the Shrinkage Limit of Soils. *ASTM Geotechnical Testing Journal*, Vol. 21, No. 3, pp. 240–250.

Chapter 14

Linear Shrinkage

INTRODUCTION

As discussed in Chapter 13, the Shrinkage Limit defines the lower limit of Water Content below which no further shrinkage of a soil–water mixture takes place. The does not really give an indication of how much shrinkage may occur; it just defines the lower bound. An alternative to the Shrinkage Limit test for evaluating shrinkage characteristics is the Linear Shrinkage test, which gives a direct measure of how much shrinkage a soil undergoes as it dries. Like the Shrinkage Limit test, since it is performed on remolded soil, the Linear Shrinkage test provides a simple index of soil behavior, in this case shrinkage potential. The test provides very good results that are very simple to obtain, does not require expensive or complicated equipment and does not use any hazardous material, e.g., mercury.

Linear Shrinkage tests (also sometimes known as *Bar Shrinkage Tests*) have been in use since at least the 1920s to provide a simple index to evaluate the shrinkage hazard of fine-grained soils (Rose 1924). Some variation of the test is performed in many countries around the world, especially in locations where abundant deposits of expansive soils are known to exist. Figure 14.1 shows two different clays that have been dried in the Linear Shrinkage mold showing different behavior.

STANDARD

At the present time there is no ASTM Standard for this test. A recommended standard procedure has been suggested by Lutenegger (2022). The Linear Shrinkage test procedure that uses the same shrinkage mold as described in this chapter is described in British Standard BS1377 and is also used in Australia and South Africa. The Texas Department of Transportation also has a similar test, called Bar Linear Shrinkage, which uses a rectangular metal shrinkage mold 19 mm × 19 mm × 127 mm (0.75 in × 0.75 in × 5.0 in).

DOI: 10.1201/9781003263289-16

Figure 14.1 Final dried length of two different clays.

EQUIPMENT

Semicircular Linear Shrinkage mold (brass, plastic or other non-corrodible metal) with internal dimensions of 140 mm ± 1 mm (5.51 in) and diameter of 25 mm ± 0.5 mm (1.0 in)

Digital vernier calipers with a capacity of 150 mm (6 in) and a resolution of 0.1 mm (0.005 in)

Silicone grease

Spatula

Ceramic or plastic mixing bowl

Electronic scale with a minimum capacity of 500 gm and a resolution of 0.01 gm

Convection drying oven capable of maintaining 110° ± 5°C

Distilled water

PROCEDURE

1. Select a representative sample of soil. The test should be performed on soil passing the No. 40 sieve. The soil may be either air-dried and pulverized or may be at natural Water Content.
2. Measure and record the initial length of the inside of a clean dry Linear Shrinkage mold using a digital caliper to the nearest 0.1 mm (L_0).

3. Apply a thin film of silicone grease to the inside of the mold.
4. Determine of the initial mass of the mold with an electronic scale to the nearest 0.01 gm (M_{MOLD}).
5. Place the soil in a clean mixing bowl. Add distilled water and thoroughly mix the soil to an initial Water Content of approximately 1.25 times the Liquid Limit of the soil (Chapter 12) (approximately 10 drops of the Casagrande Liquid Limit cup).
6. Spread the soil–water mixture into the mold in small amounts. The mixture should be worked into the mold thoroughly with a clean spatula to remove any trapped air, as shown in Figure 14.2.
7. Level off the top of the mold with a spatula and wipe the mold clean of any excess soil.
8. Determine the total mass of the mold and the soil mixture with an electronic scale to the nearest 0.01 gm ($M_{WET} + M_{MOLD}$).
9. Periodically, at intervals of approximately 2 hours, determine the length of the soil in the mold using the digital calipers, as shown in Figure 14.3. Also determine the mass of the mold and the soil.
10. Periodically, obtain measurements of length and mass until there is no further change in length. (Note: This test may take several days to complete. If the specimen is left overnight, it should be wrapped loosely with plastic wrap to prevent it from drying too fast.)
11. After the last set of measurements of total mass and length have been taken, place the mold and soil in a convection oven at 110°C for 24 h.

Figure 14.2 Adding soil to Linear Shrinkage mold.

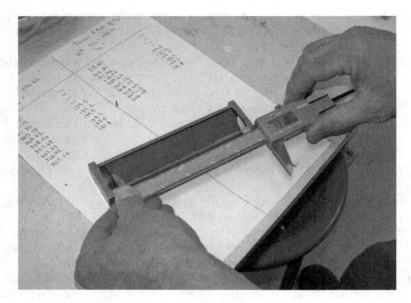

Figure 14.3 Measuring length of soil in the Linear Shrinkage mold.

12. After oven drying, remove the mold and soil from the oven and place the mold in a desiccator to cool. After cooling, obtain a final measurement of length (L_F), and mass $M_{DRY} + M_{MOLD}$) (Figure 14.4).

CALCULATIONS

Calculate Linear Shrinkage as:

$$L.S. = (L_0 - L_F)/L_0 \times 100\% \tag{14.1}$$

Use the data collected from the test and determine Water Content for each set of readings from:

$$W = \left[(M_{WET} + M_{MOLD}) - (M_{DRY} + M_{MOLD})/(M_{DRY} - M_{MOLD}) \right] \times 100\% \tag{14.2}$$

Make a plot of length vs. Water Content. Interpret the Shrinkage Limit as the intersection of the straight-line portions of the drying curve. Report the method of sample preparation (natural Water Content; air-dried; etc.); initial Water Content of the test specimen; the Linear Shrinkage to

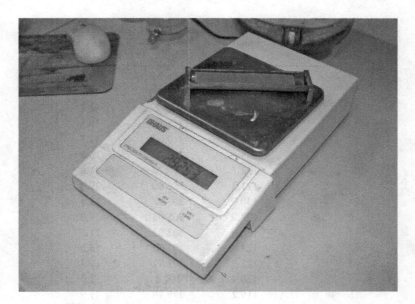

Figure 14.4 Determining final oven-dry mass of mold and soil.

the nearest 0.1%; and interpreted Shrinkage Limit to the nearest whole number.

EXAMPLE

The Linear Shrinkage data in Table 14.1 were obtained from a sample of a sample of high plasticity clay from central Missouri.

Mold No.: E   Initial Length: 140.1 mm   Initial Mold Mass: 256.56 gm

Initial Water Content = 48.5%
A plot of Water Content vs. length is shown in Figure 14.5
Linear Shrinkage: L.S. = (140.1 mm − 118.2 mm)/140.1 mm × 100% = 15.6%
Interpreted Shrinkage Limit from graph below = 9.4

DISCUSSION

According to Gilboy (1933), Atterberg actually measured the change in length of a specimen during shrinkage in order to determine the Shrinkage Limit. It is important in the Linear Shrinkage test to dry the soil very slowly. This will tend to promote more uniform shrinking along the entire length of

Table 14.1 Linear Shrinkage Data

Reading no.	Length (mm)	Wet mass (gm)	Water Content (%)
1	140.1	314.26	48.5
2	140.1	313.37	46.2
3	140.0	312.2	43.2
4	138.0	311.14	40.5
5	136.5	310.08	37.7
6	134.6	309.02	35.0
7	132.7	307.77	31.8
8	121.3	306.80	29.3
9	130.0	306.07	27.4
10	128.8	305.05	24.8
11	126.0	303.83	21.6
12	124.1	302.73	18.8
13	121.9	301.52	15.7
14	120.4	300.2	12.3
15	119.2	299.28	9.9
16	118.5	298.21	7.2
17	118.2	296.96	4.0
18	118.2	296.32	2.3
19	118.2	295.42	0.0

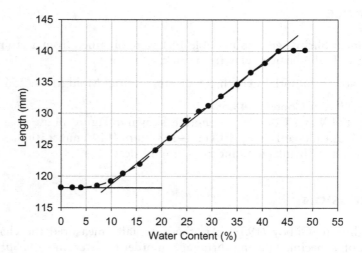

Figure 14.5 Plot of Water Content vs. length from Linear Shrinkage test.

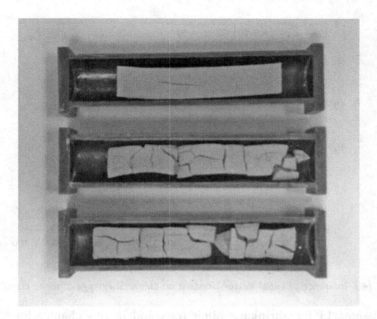

Figure 14.6 Linear Shrinkage specimens; bottom – immediately placed into oven to dry; middle – dried for a few hours and then placed in oven to dry; top – dried slowly over several days and then placed into the oven to dry.

the specimen. If the soil is dried too quickly, only a few data points will be obtained, and the soil may develop multiple cracks. Figure 14.6 shows specimens that were dried too quickly in comparison to a specimen dried slowly. Even if dried slowly, some cracks may develop in some soils. If this happens, the pieces should be slid together before obtaining a length measurement.

For some soils, the Linear Shrinkage depends on the initial Water Content, as shown in Figure 14.7 for three different clays. The lower the initial Water Content, the lower the Linear Shrinkage; the higher the initial Water Content, the higher the Linear Shrinkage. However, for soils prepared at an initial Water Content at $1.25 \times$ Liquid Limit ± 2 %W, the difference in Linear Shrinkage is rather small. Nonetheless, the initial Water Content of the specimen should always be given when reporting the Linear Shrinkage.

LINEAR SHRINKAGE BY COEFFICIENT OF LINEAR EXTENSIBILITY (COLE):

An alternative test procedure for determining the Linear Shrinkage is called the Coefficient of Linear Extensibility (COLE) test and is often used by agronomists and agricultural engineers. The COLE$_{ROD}$ test is performed by mixing the soil with water to create a paste as previously described

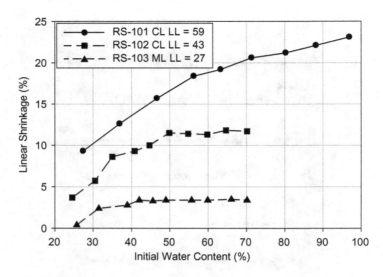

Figure 14.7 Influence of initial Water Content on Linear Shrinkage of three clays.

in Chapter 13 for Shrinkage Limit tests and in this chapter for Linear Shrinkage. The mixture is placed into the end of a plastic 25 mL syringe with a 1 cm (0.4 in) diameter orifice et the end. A soil rod with a length of about 6 cm to 10 cm (2.4 in to 4 in) is extruded. The ends of the rod are trimmed square with a moist spatula and the length is measured with a digital caliper. The rod is then allowed to air dry for 24–48 h.

After air dying, the final rod length is measured and the value of $COLE_{ROD}$ is calculated as:

$$COLE_{ROD} = \left(L_{moist} - L_{dry}\right)/L_{dry} \qquad (14.3)$$

The major differences between the $COLE_{ROD}$ test and Linear Shrinkage test are: 1) soil less than the No. 10 sieve (2 mm) is used; 2) the soil is not oven dried at the end of the test; 3) determination of initial and final Water Content is optional; and 4) only the % Linear Shrinkage is obtained.

The $COLE_{ROD}$ test originated from the $COLE_{CLOD}$ test, which is performed on undisturbed samples. Additional details of the test and correlations to other soil properties are given by Schafer and Singer (1976) and DeJong et al. (1992).

REFERENCES

DeJong, E., Kozak, L.M., and Stonehouse, H.B., 1992. Comparison of Shrink-Swell Indices of Some Saskatchewan Soils and Their Relationships to Standard Soil Characteristics. *Canadian Journal of Soil Science*, Vol. 72, pp. 429–439.

Gilboy, G., 1933. Soil Mechanics Research. *Transactions of the ASCE*, Vol. 98, pp. 221–236.

Lutenegger, A.J., 2022. Recommended Method for Performing Linear Shrinkage Test. Submitted to ASTM Geotechnical Testing Journal.

Rose, A.C., 1924. Practical Field Tests for Subgrade Soils. *Public Roads*, Vol. 5, No. 6, pp. 10–15.

Schafer, W.M., and Singer, M.J., 1976. A New Method of Measuring Shrink-Swell Potential Using Soil Pastes. *Soil Science Society of America Journal*, Vol. 40, pp. 805–806.

ADDITIONAL READING

Anderson, J.U., Fadul, K.E., and O'Connor, G.A., 1973. Factors Affecting the Coefficient of Linear Extensibility in Vertisols. *Soil Science Society of America Proceedings*, Vol. 37, pp. 296–299.

Delaney, M.D., Allman, M.A., Smith, D.W., and Sloan, S.W., 1998. The Establishment and Monitoring of Expansive Soil Field Sites. *Geotechnical Site Characterization*, pp. 551–558.

Grossman, R.B., Brasher, B.R., Franzmeier, D.P., and Walker, J.L., 1968. Linear Extensibility as Calculated from Natural Clad Bulk Density Measurements. *Soil Science Society of America Proceedings*, Vol. 32, pp. 570–573.

Heidema, P.B., 1957. The Bar-Linear Shrinkage Test and the Practical Importance of Bar-Linear Shrinkage as an Identifier of Soils. *Proceedings of the 4th International Conference on Soil Mechanics and Foundation Engineering*, Vol. 1, pp. 44–49.

Hobbs, P.R., Northmore, K.J. Jonmes, L.D., and Entwisle, D.C., 2000. Shrinkage Behaviour of Some Tropical Clays. *Unsaturated Soils for Asia*, pp. 675–680.

Jackson, J.O., 1981. Geotechnical Properties of the Biu Black Cotton Soil. *Proceedings of the 10th International Conference on Soil Mechanics and Foundation Engineering*, Vol. 3, pp. 641–644.

Page-Green, P., and Ventura, D., 1999. The Bar Linear Shrinkage Test – More Useful Than We Think. *Geotechnics for Developing Africa*, pp. 379–387.

Ring, G.W., 1966. Shrink-Swell Potential of Soils. *Highway Research Record*, Vol. 119, pp. 17–21.

Tadanier, R., and Nguyen, V.U., 1984. Index Properties of Expansive Soils in New South Wales. *Proceedings of the 5th International Conference on Expansive Soils*, pp. 321–328.

Wall, G.D., 1959. Observations of the Use of Lineal Shrinkage from Liquid Limit as an Aid to Control in the Field. *Proceedings of the 2nd African Regional Conference on Soil Mechanics and Foundation Engineering*, pp. 175–180.

Widjaja, B., and Chriswandi, C., 2020. New Relationship Between Linear Shrinkage and Shrinkage Limit for Expansive Soils. *Proceedings of the 3rd TICATE IOP Conference Series: Materials Science and Engineering*.

Chapter 15

Shrink Test

INTRODUCTION

As noted in previous chapters, most fine-grained soils tend to shrink when water is removed. For example, during a long drought, the ground will often have cracks showing at the surface, or large trees around a house can draw water from the soil, causing the house to settle as the soil shrinks. These same soils also tend to swell when water is added, causing other problems. Unlike the Shrinkage Limit test and the Linear Shrinkage test, which are really index tests performed on remolded soils at high Water Content, the Shrink Test determines the shrinkage behavior of soils with their natural fabric, and therefore the results are more representative of the behavior that would occur in the field.

A Shrink Test may be used to directly determine how much shrinkage will take place as a soil dries. Tests may be performed on natural core samples or on compacted samples trimmed from a sample after being removed from the compaction mold. These tests are relatively simple to perform and involve periodically measuring the volume and mass of a specimen as drying is allowed over a period of several days. When performed on natural soil cores, Shrink Tests provide a direct measurement of how the natural undisturbed soil will behave.

STANDARD

There currently is no ASTM standard test procedure for the Shrink Test. A procedure was presented by Briaud et al. (2003) and is the procedure described in this chapter. A similar test was described by Hanafy (1998) using the ring from a one-dimensional consolidation cell. British Standard test BS 1377 – Part 2 describes a similar method.

DOI: 10.1201/9781003263289-17

EQUIPMENT

Spatula or trimming knife
Small glass or plexiglass plates approximately 100 mm × 100 mm (4 in × 4 in)
Ceramic bowls
Digital vernier calipers with a capacity of 150 mm (6 in) and a resolution of 0.1 mm (0.005 in)
Electronic scale with a minimum capacity of 600 gm and a resolution of 0.01 gm.
Convection drying oven capable of maintaining 110° ± 5°C.

PROCEDURE

1. Trim a specimen from a core sample or compacted sample by using a spatula or knife to square off the ends. Trim the specimen so that the height is about 20–40 mm (0.8 to 1.5 in). Square off the ends of the specimen.
2. Using the digital vernier calipers, measure the length and diameter of the specimen to the nearest 0.1 mm as shown in Figures 15.1 and 15.2. (Note: If a cubical specimen is being used, trim off all sides and measure the height, width and length of the specimen.) Obtain at least three measurements of each dimension and record all three

Figure 15.1 Measurement of the diameter of a trimmed specimen.

Figure 15.2 Measurement of the length of a trimmed specimen.

measurements. Use the average of each measurement and calculate the total volume, V_T, of the specimen.

$$V_T = \pi r^2 L \quad \text{(for cylindrical specimens)}$$

$$V_T = L \times W \times H \quad \text{(for cubical specimens)}$$

3. Determine the mass of a clean dry glass or plexiglass plate (M_{PLATE}) to the nearest 0.01 gm. Place the trimmed specimen on the plate and determine the total mass ($M_{WET} + M_{PLATE}$), to the nearest 0.01 gm.
4. Place the specimen on the plate so that the specimen rests on the diameter, not flat.
5. Periodically, at intervals of about 2–3 hours, take measurements of the length, diameter and mass of the specimen. (Note: Do not leave the specimen in sunlight and wrap the specimen in plastic wrap overnight to prevent rapid drying.)
6. Continue taking measurements until the length and diameter of the specimen show no further change. (Note: The specimen can generally be placed on its side so that drying will be more uniform. An important part of this test is to dry the specimen very slowly. This may take from 4 to 7 days depending on the soil.)
7. Determine the mass of a clean dry ceramic bowl (M_{BOWL}). Place the specimen in the bowl and determine the total mass ($M_{WET} + M_{BOWL}$). Now place the bowl in a convection oven at 110°C for 24 hours.

8. After 24 hours, remove the bowl and specimen from the oven and place the soil and bowl in a desiccator to cool. After cooling, determine the oven dry mass of the soil and bowl $(M_{DRY} + M_{BOWL})$. Measure the final diameter and final length of the oven-dried specimen.

CALCULATIONS

Calculate the initial volume of the specimen, V_0.
Calculate the initial Water Content of the specimen from:

$$w_0 = \left[(M_{WET} + M_{BOWL}) - (M_{DRY} + M_{BOWL})/(M_{DRY} - M_{BOWL})\right] \times 100\%$$

For each set of measurements, calculate the volume using the average diameter and length of the specimen and calculate the Water Content of the specimen using the final dry mass of the soil.

Present the data in a plot of volume vs. Water Content. The limit of shrinkage is interpreted from the graph as the intersection of two straight line segments.

Using the volume and Water Content data from each set of readings, make a plot of cumulative ΔW (decimal) vs. cumulative $\Delta V/V_0$ (% in decimal).

Calculate the shrink modulus from the straight line portion of the curve as:

$$E_w = \Delta w / (\Delta V/V_0)$$

EXAMPLE

Data from a Shrink Test on an undisturbed sample of clay till from Winterset, Iowa, are given in Table 15.1.

Table 15.1 Data from Shrink Test

Reading no.	Mass (gm)	Average diameter (mm)	Average length (mm)	Volume (cm³)	Water Content (%)
1	122.91	63.4	20.1	63.5	25.6
2	118.87	62.6	19.5	60.0	21.4
3	115.51	61.5	19.4	57.6	18.0
4	111.41	60.4	19.1	54.7	13.8
5	107.67	59.8	18.9	53.1	10.0
6	104.54	59.4	18.8	52.1	6.8
7	101.01	59.31	18.7	51.6	3.2
8	97.88	59.21	18.7	51.5	0.0

Initial wet mass: 122.91 gm; initial diameter: 63.4 mm; initial length: 20.1 mm

From the graph shown in Figure 15.3, the interpreted Limit of Shrinkage is 9.4%.

From the plot of $\Delta V/V_0$ vs. ΔW shown in Figure 15.4, $E_w = 1.31$.

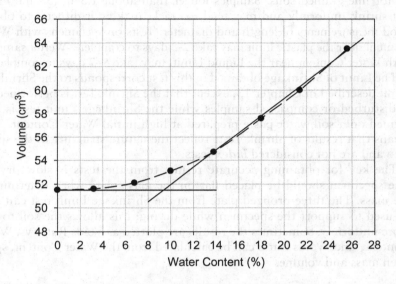

Figure 15.3 Plot of Water Content vs. volume from Shrink Test.

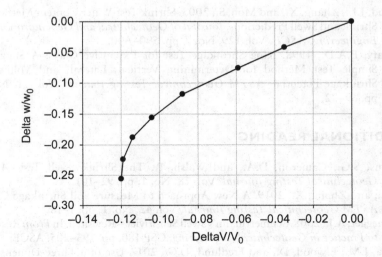

Figure 15.4 Plot of $\Delta V/V_0$ vs. ΔW from Shrink Test.

DISCUSSION

The Shrink Test is a simple test to perform and gives an inexpensive direct measurement of the shrinkage behavior of soils. The test requires no special laboratory equipment other than a digital caliper, which is used to reduce errors in measurements of length and diameter. The test is suited primarily for fine-grained soils. Samples longer than about 1.5 in (38 mm) may not shrink uniformly and may tend to crack, making it difficult to obtain good measurements of length and diameter. Tests on specimens with Water Content near the plastic limit may take 2-3 days to complete. Wetter samples, with Water Content near the Liquid Limit, may take 5–7 days to complete.

The Limit of Shrinkage obtained in this test corresponds to the Shrinkage Limit described in Chapter 13, except that the Shrink Test is performed on undisturbed or compacted samples while the Shrinkage Limit test is performed on a soil:water paste prepared at high initial Water Content. This means that results of Shrink Tests reflect the nature structure of the specimen and are not considered *index* tests.

The key for obtaining accurate results from the tests is slow drying. The specimen should be placed flat on the glass plate when determining the mass. The three-pronged plate from the Shrinkage Limit test can also be used to support the specimen while drying; this allows the soil to dry more uniformly. Sometimes the results are plotted as Voids Ratio vs. Water Content, since Voids Ratio can be calculated from the Water Content, specimen mass and volume.

REFERENCES

Briaud, J.L., Zhang, X., and Mon, S., 2003. Shrink Test-Water Content Method for Shrink and Swell Predictions. *Journal of Geotechnical and Geoenvironmental Engineering, ASCE*, Vol. 129, No. 7, pp. 590–600.
Hanafy, E.A.D., 1998. Ring Shrinkage Test for Expansive Clays: A Suggested Simple Test Method for Determining Vertical, Lateral, and Volumetric Shrinkage Potential. *ASTM Geotechnical Testing Journal*, Vol. 21, No. 1, pp. 69–72.

ADDITIONAL READING

Frityus, S.G., Cameron, D.A., and Walsh, P., The Shrink Swell Test. *ASTM Geotechnical Testing Journal*, Vol. 28, No. 1, pp. 92–101.
Li, L., and Zhang, X., 2019. A New Approach to Measure Soil Shrinkage Curve. *ASTM Geotechnical Testing Journal*, Vol. 42, No. 1, pp. 1–18.
Lutenegger, A.J., 2008. Schmertmann's Swell Sensitivity – Revisited. In *From Research to Practice in Geotechnical Engineering*, GSP 180, pp. 193–205. ASCE.
Wong, J.M., Elwood, D., and Fredlund, D.G., 2017. Use of a Three-Dimensional Scanner for Shrinkage Curve Tests. *Canadian Geotechnical Journal*, Vol. 56, pp. 526–535.

Chapter 16

Free-Swell Index Tests

BACKGROUND

Expansive (swelling) soils are those soils that experience volume increase (swell) when water is added or volume decrease (shrink) when water is removed, as by drying. Generally, if a given soil shows high shrink behavior, it will also generally show high swell behavior, depending on the initial Water Content. These soils are often referred to as "high shrink-swell soils". In general, these soils experience such volume changes because of the presence of expansive clay minerals belonging to the smectite (montmorillonite) group of clay minerals. Simple tests that can be used to help identify soils that may be prone to exhibit detrimental swelling behavior are very useful to the engineer. Detailed maps are also often available for certain geographic areas showing the extent of known expansive soils. The effects of shrinking and swelling in an area can often be seen as a "stair-step" crack pattern on concrete block or brick walls, as shown in Figure 16.1.

Most expansive soils have low natural Water Content (often close to or below the Plastic Limit) and typically have high natural dry density. The Liquidity Index is often close to zero or may even be negative. Figure 16.2 shows the variation in natural Water Content and Atterberg Liquid Limit, Plastic Limit and Shrinkage Limit versus depth in a saturated expansive soil profile from central Missouri. Note that in this case, the natural Water Content is very close to the Plastic Index, suggesting that the Liquidity Index is near zero. The soil could still shrink down to the Shrinkage Limit if it is exposed to drying, but it doesn't have far to go to get to the Shrinkage Limit. However, if the soil becomes wetted, as from a broken water pipe or excessive rainfall and infiltration, it has a long way to go to get to the Liquid Limit, even though it probably would never make it that far.

DOI: 10.1201/9781003263289-18

Figure 16.1 Typical wall crack patterns observed in structures built on expansive soils.

FREE-SWELL INDEX (FSI)

Introduction

The Free-Swell Index Test is a simple test that is used to indicate whether a soil has high shrink–swell tendency. The test was developed in the 1950s by Holtz and Gibbs (1956) and involves placing a small amount of soil in distilled water and watching to see how much the soil increases in volume. the Free-Swell Index Test is a direct swell test that gives a qualitative indication of the severity of swell potential through the quantitative measurement of swelled volume, as compared to non-swelled volume. The results can be used to compare soils and should be used as an indicator test showing whether more sophisticated direct tests should be performed on a soil to quantify the swelling behavior under certain conditions of confinement and Water Content change.

Since the test is performed on soils that have been dried, crushed and sieved through the No. 40 sieve, the results are only influenced by soil composition, such as clay content, clay mineralogy, Cation Exchange Capacity and Specific Surface Area. This means that the results can be compared from one soil to the next, and, based on the test results, we can classify the soil in terms of its potential degree of swelling severity.

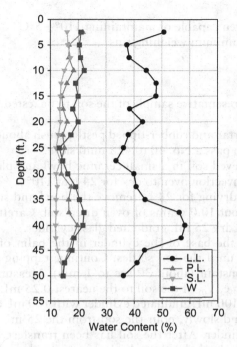

Figure 16.2 Atterberg Limits and Water Content profile of an expansive soil in Central Missouri.

Standard

There is currently no ASTM standard procedure for this test described in this chapter. ASTM D5890 *Standard Test Method for Swell Index of Clay Mineral Component of Geosynthetic Clay Liners* describes a similar test procedure using 2 gm of soil for testing bentonite in conjunction with the use of geosynthetic clay liners. The test procedures described in this chapter are based on the suggestion of Holtz and Gibbs (1956) as modified by Shridharan et al. (1985) and Sivapullaiah et al. (1987). Lutenegger (2022) has proposed a recommended standard test procedure.

Equipment

   No. 40 sieve
   Small ceramic bowls
   25 mL graduated glass cylinder
   100 mL graduated glass cylinder
   Ceramic mortar and rubber-tipped pestle
   Electronic scale with a minimum capacity of 250 gm and a resolution of
      0.01 gm

Distilled water
Convection oven capable of maintaining $110° \pm 5°C$
Ceramic or aluminum weighing tins

Procedure

1. Select a representative sample of the soil to be tested. Allow the soil to air dry.
2. Using a mortar and rubber-tipped pestle, crush about 50 grams of air-dried soil to pass a No. 40 (0.425 mm) sieve.
3. Place the sieved soil in a small ceramic bowl and place the bowl and soil in a convection oven to dry for 24 h at 110°C.
4. After oven drying for 24 h, remove the bowl and soil from the oven and weigh out 10.0 grams of oven dried soil. Carefully pour the soil into a clean dry 25 mL graduated glass cylinder.
5. Gently tap the base of the cylinder in the palm of the hand or on a table top until the soil settles. Continue tapping until the volume becomes constant (about 30 sec to 1 min). Measure and record the initial volume, V_0, of the soil to the nearest 0.25 mL.
6. Fill a clean 100 mL graduated cylinder with 100 mL of distilled water. Carefully and slowly, pour the soil from the 25 mL cylinder into the 100 mL cylinder. After the soil has been transferred to the 100 mL cylinder, gently shake the cylinder or stir with a clean glass rod to remove any trapped air pockets and any soil adhering to the sides of the cylinder. Set the cylinder aside to rest for 24 hours.
7. After 24 hours, read the final volume, V_F, from the markings on the graduated cylinder to the nearest 0.25 mL. Interpolate if necessary. Figure 16.3 shows swelled volume of two soils (with the same dry mass of 10 gm) after 24 h (Figure 16.4).

Calculations

Calculate the Free-Swell Index, FSI, as:

$$FSI = \left[(V_F - V_0) / V_0 \right] \times 100\% \tag{16.1}$$

Report FSI to nearest whole percentage.

Calculate the Modified Free-Swell Index, MFSI, as:

$$MFSI = \left[(V_F - V_S)/V_S \right] \tag{16.2}$$

where:
   V_S = volume of solids = M_{DRY}/ρ_s) (Note: Use measured value of ρ_s if available, otherwise assume $\rho_s = 2.65$ gm/cm^3)

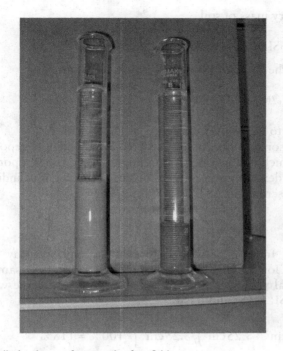

Figure 16.3 swelled volume of two soils after 24 h.

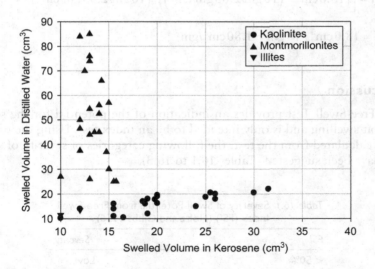

Figure 16.4 Swelled volume of some pure clay minerals in water and kerosene.

M_{DRY} = dry mass of soil

Report MFSI to nearest 0.1

Calculate the Relative Free-Swell Index, RFSI, as:

$$RFSI = V_F/M_{DRY} \tag{16.3}$$

Report RFSI to nearest 0.1 cm³/gm

(Note: In some publications, the Free-Swell Index is reported in units of mL/gm which is the same as cm³/gm; however, other reported values of Free-Swell Index are given in units of mL or cm³/2gm to indicate that the procedure described in ASTM D5890 has been used.)

Example

A Free-Swell Index Test was performed on a plastic clay sample from Southeastern Iowa (NL-YSP). The oven dried mass of the sample was 10.0 gm, the initial volume was 8.25 cm³ and the final volume was 18.0 cm³. Determine FSI, MFSI and RFSI.

$$FSI = \left[\left(18.0 \, cm^3 - 8.25 \, cm^3 \right) \big/ 8.25 \, cm^3 \right] \times 100\% = 118\%$$

$$MFSI = \left[\left(18.0 \, cm^3 - \left(10 \, gm/2.65 \, gm/cm^3 \right) \right) \big/ \left(10 \, gm/2.65 \, gm/cm^3 \right) \right] = 3.8$$

$$RFSI = 18.0 \, cm^3 \, / \, 10 \, gm = 1.80 \, cm^3/gm$$

Discussion

The Free-Swell Test provides an indication of the potential for the soil to exhibit swelling and is only intended to be an index test. Using the parameters calculated from the test, the following categories of severity of swelling have been suggested (Tables 16.1 to 16.3).

Table 16.1 Severity of Swell Potential from Free-Swell Index (FSI) (Holtz and Gibbs 1956)

FSI	Severity
< 50%	Low
50–100%	Medium
100–140%	High
> 140%	Very high

Table 16.2 Severity of Swell Potential from Modified
Free-Swell Index (MFSI) (Sivapullaiah et al. 1987)

MFSI	Severity
< 2.5	Negligible
2.5–10.0	Moderate
10.0–20.0	High
> 20.0	Very high

Table 16.3 Severity of Swell Potential from Relative
Free-Swell Index (RFSI) (Sridharan et al. 1985)

RFSI	Severity
< 1.5	Negligible
1.5–2.0	Slight
2.0–5.0	Moderate
5.0–10.0	High
> 10.0	Very high

DIFFERENTIAL FREE-SWELL INDEX (DFSI) AND FREE-SWELL RATIO (FSR)

Introduction

Another modification that has been made to the Free-Swell Index Test is to compare the swelling behavior of soils in both a polar and non-polar fluid, such as water and kerosene. This procedure appears to have become a standard method for identifying soils in India and some other parts of the world that have large amounts of swelling soils.

Standard

There is currently no ASTM standard test procedure for this test.

Equipment

No. 40 sieve
Small ceramic bowls
25 mL graduated glass cylinder
100 mL graduated glass cylinder
Ceramic mortar and rubber-tipped pestle
Electronic scale with a minimum capacity of 250 gm and a resolution of 0.01 gm
High grade kerosene
Convection oven capable of maintaining $110° \pm 5°C$
Ceramic or aluminum weighing tins

Procedure

1. Select a representative sample of soil and perform the Free-Swell Index Test as described in Section 16.1.
2. Carefully weigh out a duplicate 10 gm sample of oven-dried soil as used above.
3. Fill a 100 mL graduated cylinder with 40 mL of kerosene. Pour the soil sample from the 25 mL graduated cylinder into the 100 mL graduated cylinder. Fill the 100 mL cylinder to the 100 mL mark with kerosene. Gently stir with a clean glass rod to remove any trapped air pockets. Set the cylinder aside to rest for 24 h.
4. After 24 h, read the final volume, V_K, to the nearest 0.25 mL from the markings on the graduated cylinder.
5. At the end of the test, dispose of the soil and kerosene in appropriate containers.

Calculations

Calculate the Differential Free-Swell Index, DFSI, as:

$$DFSI = \left[\left(V_W - V_K \right) / V_K \right] \times 100\% \qquad (16.4)$$

where:
V_W = volume of soil in water
V_K = volume of soil in kerosene

Report the DFSI to the nearest whole percent.
Calculate the Free-Swell Ratio, FSR, as:

$$FSR = \left(V_W \right) / \left(V_K \right) \qquad (16.5)$$

Report the FSR to the nearest 0.1.

(Note: Some publications use a different definition for the term *Free-Swell Index*, which can cause some confusion. Instead of using Eq. 16.1, FSI is defined in terms of the swelling in water and kerosene and according to Eq. 16.4. Readers should make sure that when reading publications in the literature which definition of FSI has been used.)

Example

The plastic clay sample from Southeastern Iowa (NL-YSP) tested in Section 16.1 was also tested to determine the Differential Free-Swell Index and the Free-Swell Ratio. The oven-dried mass of the sample was 10.0 gm, and the final volume of a duplicate sample in kerosene was 10.0 cm³. Determine DFSI and FSR.

$$DFSI = (18.0 \, cm^3 \quad 10.0 \, cm^3 / 10.0 \, cm^3 \times 100\% = 80\%$$

$$FSR = \left(18.0\,\text{cm}^3\right)/\left(10.0\,\text{cm}^3\right) = 1.8$$

Discussion

The swelling test in kerosene is simple to perform, but remember that the waste kerosene must be disposed of properly at the end of the test. Values of DFSI and FSR may be used to help identify swelling soils based on Tables 16.4 and 16.5.

Test results presented by Lutenegger (2019) indicate that the final volume of soil measured in water and kerosene may be used to give an indication of clay mineralogy. Figure 16.5 shows results obtained using some pure clay minerals. These test results indicate Free-Swell Ratios for different clay minerals as:

Kaolinite: FSR = 0.6 – 1.0
Montmorillonite: FSR = 5 – 8
Illite: FSR = 1.0 – 1.5

The Free-Swell Index is a measure of how much a fixed amount of soil will swell in water, and therefore, since there no other factors are involved, the FSI should be related only to surface characteristics of the soil., i.e., CEC and SSA. As an example, Figure 16.5 shows the results of Free-Swell Tests performed by the author on several clays suggesting an influence of SSA on FSI. The data are not universal since the soils come from a number of locations around the world and have different geologic origins. Other factors,

Table 16.4 Expansion Potential of Soils Based on Differential Free-Swell Index (Mohan 1977)

DFSI	Expansion Potential
< 50	Low
50–100	Medium
100–200	High
> 200	Very high

Table 16.5 Expansion Potential of Soils Based on Free-Swell Ratio (Prakash & Sridharan 2004)

FSR	Expansion potential
< 1.0	Negligible
1.0–1.5	Low
1.5–2.0	Moderate
2.0–4.0	High
> 4.0	Very high

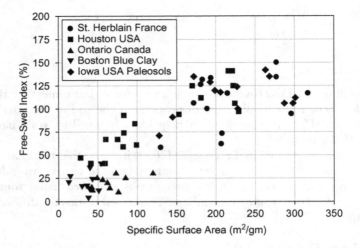

Figure 16.5 Influence of Specific Surface Area on Free-Swell Index.

such as Carbonate Content or original pore fluid composition, may influence FSI but are not accounted for in the measurement of SSA.

The Free-Swell Index can also be used as an indicator test to evaluate the influence of different pore fluids on the swelling behavior or the potential effectiveness of soil stabilizers on swelling. Several studies have shown that the Free-Swell Index may be used as a simple test to estimate the swelling pressure or swell strain of undisturbed or compacted clays (e.g., Jayasekera & Mohajerani 2003; Rao et al. 2004; Patil et al. 2016; Radwar & Bahloul 2019). For example, Figure 16.6 shows test results performed by the author on two clays and the influence of quicklime (CaO) on the swelling behavior.

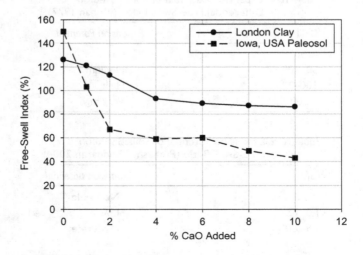

Figure 16.6 Influence of quicklime (CaO) on Free-Swell Index of two clays.

REFERENCES

Holtz, W.G., and Gibbs, H.J., 1956. Engineering Properties of Expansive Clays. *Transactions of the American Society of Civil Engineers*, Vol. 121, pp. 641–663.

Jeyaskera, S., and Mehajerani, A., 2003. Some relationships between Shrink-Swell Index, Liquid Limit, Plasticity Index, Activity and Free-Swell Index. *Australian Geomechanics*, Vol. 38, No. 2, pp. 53–58.

Lutenegger, A.J., 2019. *Soils and Geotechnology in Construction*. Taylor & Francis Publishers.

Lutenegger, A.J., 2022. Recommended Test Procedure for Performing the Free-Swell Index Test. Submitted to ASTM Geotechnical Testing Journal.

Mohan, D., 1977. Engineering of Expansive Soils. *Proceedings of the 1st National Symposium on Expansive Soils*, Kanpur, India.

Patil, A.P., Parkhe, D.D, Shrigriwar, R.V., and Panse, R.V., 2016. Establishing Relationships Between Swelling Pressure and Free Swell Index of Soils – A Case Study. *Proceedings of the National Conference on Recent Innovations in Science Engineering and Technology*, pp. 4–7.

Prakash, K., and Sridharan, A., 2004. Free Swell Ratio and Clay Mineralogy of Fine-Grained Soils. *Geotechnical Testing Journal, ASTM*, Vol. 27, No. 2, pp. 220–225.

Radwar, N.A., and Bahloul, K.M., 2019. Correlation Between swelling Pressure and Free Swell Index of Greater Cairo City Expansive Soils – A Case Study. *Journal of Engineering Research and Reports*, Vol. 6, No. 3, pp. 1–6.

Rao, A.S., Phanikumar, B.R., and Sharma, R.S., 2004. Prediction of Swelling Characteristics of Remoulded and Compacted Expansive Soils Using Free Swell Index. *Quarterly Journal of Engineering Geology and Hydrogeology*, Vol. 37, pp. 217–226.

Sivapullaiah, P.V., Sitharam, T.G., and Subba Rao, K.S., 1987. Modified Free Swell Index for Clays. *Geotechnical Testing Journal, ASTM*, Vol. 10, No. 2, pp. 80–85.

Sridharan, A., Rao, S.M., and Murthy, N.S., 1985. Free-Swell Index of Soils: A Need for Redefinition. *Indian Geotechnical Journal*, Vol. 15, No. 2, pp. 80–85.

ADDITIONAL READING

El-Sohby, M.A., Mazen, S.O., and Abdil, M.T., 1988. Evaluation of Free-Swell Test Measurements. *Proceedings of the International Conference on Engineering Properties of Regional Soils*, pp. 568–571.

Sridharan, A., and Prakash, K., 2000. Classification Procedures for Expansive Soils. *Proceedings of the Institution of Civil Engineers – Geotechnical Engineering*, Vol. 143, pp. 235–240.

Sridharan, A., and Rao, S.M., 1988. A Scientific Basis for the Use of Index Tests in Identification of Expansive Soils. *ASTM Geotechnical Testing Journal*, Vol. 11, No. 3, pp. 208–212.

Sridharan, A., Rao, S.M., and Murthy, N.S., 1986. A Rapid Method to Identify Clay Type in Soils by the Free-Swell Technique. *ASTM Geotechnical Testing Journal*, Vol. 9, No. 4, pp. 198–203.

Sridharan, A., Rao, S.M., and Joshi, S., 1990. Classification of Expansive Soils by Sediment Volume Method. *ASTM Geotechnical Testing Journal*, Vol. 13, No. 4, pp. 375–380.

Stamatopoulos, A.C., Christodoulias, J.C., and Giannoros, H.C., 1992. Treatment of Expansive Soils for Reducing Sell Potential and Increasing Strength. *Quarterly Journal of Engineering Geology*, Vol. 25, pp. 301–312.

Chapter 17

Dispersion

BACKGROUND

Dispersive soils are those which normally deflocculate or are easily dis-aggregated when exposed to water of low salt concentration. According to Bell and Maud (1994), Dispersion occurs in soils when the repulsive forces between clay particles exceed the attractive forces. This brings about deflocculation, so that when exposed to relatively pure water, the particles repel each other. This is actually the basis for adding a "dispersing agent" to the soil suspension when conducting the Hydrometer or Pipette Test to determine the grain-size distribution of fine-grained soils.

In low salinity or distilled water, the cohesion between soil particles is partially or completely lost, and soil aggregates are dispersed into individual particles. The behavior of dispersive soils is complex and is related to the type of clay minerals present, the existence of water soluble cementing agents and cation exchange. In the field, dispersive soils often can be traced to the cause of failures of compacted embankments and earth dams by the formation of external gullies and holes and by internal open channels developed within the earth structure.

Recall that in Chapter 4 dispersing agent had to be added to soils in order to correctly determine the grain-size distribution of the finer fraction of the sol using either the Hydrometer Test or the Pipette Test. The "dispersing agent" changes the chemistry of the fluid and produces deflucculation so that the true grain-size distribution can be measured. The dispersing agent (and agitation) breaks down the aggregates to find the correct amount of each of the grain sizes. In effect, Dispersion is an indication of soil aggregate stability.

There are three standard test procedures available for evaluating the dispersive characteristics (sometimes referred to as *dispersivity*) of fine-grained soils: 1) Double Hydrometer test; 2) Pinhole Dispersion Test; and 3) Crumb Test. Results of these tests give an indication of the erosivity of a soil, which may be important in the design of earth dams and canals, the protection of river banks, scour potential and sediment load in rivers and streams.

DOI: 10.1201/9781003263289-19

DISPERSION BY DOUBLE HYDROMETER TEST

Introduction

In this test, developed in the 1940s by the U.S. Soil Conservation Service (Sherard et al. 1976; 1976a; Decker & Dunnigan 1977), a standard Hydrometer Test is performed with dispersing agent and mixing. The test is then performed on an identical soil specimen but in this case no dispersing agent is used and there is no agitation. The degree of Dispersion is defined by comparing the percentage of soil particles < 0.005 mm in the test with dispersing agent to the percent < 0.005 mm without dispersing agent.

Standard

ASTM D4221 *Standard Test Method for Dispersive Characteristics of Clay Soil by Double Hydrometer*

Equipment

No. 40 sieve
Electronic scale with a minimum capacity of 600 gm and a resolution of 0.01 gm
1000 mL hydrometer cylinders
152H Hydrometer
Thermometer with a range of 0 to 50°C and a resolution of 0.2°C
500 mL beakers
Distilled water
Plastic wash bottles
Rubber stoppers to fit 1000 mL hydrometer cylinders
Stopwatch

Procedure

1. Obtain the equivalent of approximately 50 g of oven-dry soil passing the No. 40 sieve. Use the same soil that was used in the standard Hydrometer Test. (Note: do not oven dry the soil; use $M_{WET} = M_{DRY} \times (1 + W/100)$ and take a separate sample to determine the Water Content.)
2. Mix the soil to a thick slurry in a 500 mL beaker using 125 mL of distilled-deionized water. Stir until the soil is thoroughly wetted.
3. Allow the solution to soak for a minimum of 16 hours.
4. Prepare a reference liquid of distilled water.
5. Transfer the slurry to a 1000 mL hydrometer cylinder and add distilled water as necessary to bring the total volume to 1 liter.
6. Place your palm over the end of the cylinder and mix the solution for 1 minute by turning the cylinder upside down and back. Set the cylinder down and simultaneously start a timer (t = 0).

7. Obtain enough hydrometer readings to determine the percent of material finer than 0.005 mm. The hydrometer should be read at the top of the meniscus, and after each reading it should be removed from the suspension, rinsed and placed in the reference liquid between readings. Also, the cylinder with the soil suspension should be covered with a glass plate between readings.

8. Correct the hydrometer readings for temperature and the meniscus by taking a reading in the reference solution. Obtain this correction each time a reading of the sample is taken. Make a separate determination of the meniscus correction using distilled water.

Calculations

For the Hydrometer Test, the maximum diameter still in solution at and/or above the center of volume of the hydrometer is computed as:

$$D = K \times (L/T)^{0.5} \tag{17.1}$$

where:

D = particle diameter, mm
L = effective depth, cm (Table 2, ASTM D422, use original hydrometer reading corrected only for the meniscus; see Appendix II)
T = elapsed time, minutes
K = constant (Table 3, ASTM D422; see Appendix II)
For hydrometer 152H the percent passing, P, is computed as:

$$P(\%) = (Ra/M) \times 100\% \tag{17.2}$$

where:

R = corrected hydrometer reading (actual reading – control fluid reading)
M = oven-dry mass of soil
a = correction factor (Table 1, ASTM D422; see Appendix II)

Calculate the % Dispersion as:

$$\text{Dispersion} = (A/B) \times 100\% \tag{17.3}$$

where:

A = % < 0.005 mm from "non-dispersed" test
B = % < 0.005 mm from standard test

Report Dispersion to the nearest 0.1%.

Example

A sample of clay from Salt Lake City, Utah, was tested in a standard Hydrometer Test and gave % < 0.005 mm = 40.4%.

The same soil was tested without dispersing agent and without agitation and gave % < 0.005 mm = 11.9%.

Calculate the Dispersion.

$$\text{Dispersion} = (11.9 / 40.4) \times 100\% = 29.5\%$$

Discussion

From Equation 17.3, if the soil is easily "dispersed" in just plain water and gives the same amount of < 0.005 mm as a standard test performed with added dispersing agent and agitation, the soil is considered highly dispersive. That is, if Dispersion = 100%, the soil is completely dispersive; if Dispersion = 0%, the soil is completely non-dispersive. ASTM D4221 indicates that about 85% of dispersive clays exhibit a Dispersion according to Equation 17.3 of greater than 35%. The test could also be performed using the pipette method. The soil is described as non-dispersive if the Dispersion ratio is less than 15%; *slightly dispersive* if the ratio is between 15% and 30%; *moderately dispersive* if the Dispersion ratio is between 30% and 50%; and *highly dispersive* if the Dispersion ratio is greater than 50% (Elges 1985).

Dispersion tests are a way to determine how easily a soil might become disaggregated. The tests have real consequences, since it may be important to determine if a soil has a tendency to disaggregate easily when it comes into contact with clean water. In natural or compacted soils that may be subjected to water flow, dispersive soils tend to create a condition known as "piping", which is a form of internal erosion. Piping tends to produce internal flow cavities by washing away soils such as in an earth dam or in backfill around buried pipes. The Double Hydrometer Test is an indicator test giving results that suggest more testing may be needed. The test could also be performed using the pipette method.

DISPERSION BY PINHOLE TEST

Introduction

Another test for identifying dispersive soils, especially compacted soils, was also developed by the U.S. Soil Conservation Service and is called the pinhole test, and was described by Sherard et al. (1976). In this test, distilled water is allowed to flow under a small gradient through a 1.0 mm (0.04 in) diameter hole in a compacted soil sample contained in a simple test chamber, as shown in Figure 17.1. Dispersive soils will rapidly erode soil as water

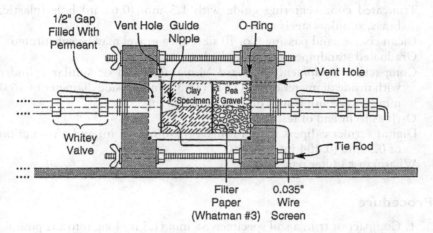

Figure 17.1 Schematic of pinhole test for identifying dispersive soils (after ASTM D4647).

flows through the hole under a small pressure head. A rapid increase in the hole size is indicated by an increase in flow rate and increased turbidity of the effluent. For dispersive soils, the water on the outflow end of the sample becomes cloudy and the hole rapidly erodes. For non-dispersive soils, there is generally only a small amount of erosion, and water tends to stay clear.

A similar test was described by Lefebvre et al. (1985). The "drill hole test" was developed primarily to evaluate the erosivity of intact clays and minimizes the surface disturbance and allows the measurement of tractive forces.

Standard

ASTM D4647 *Standard Test Method for Identification and Classification of Dispersive Soils by the Pinhole Test*

Equipment

Pinhole test apparatus
Constant head tank
Distilled water
Electronic scale with a minimum capacity of 1000 gm and a resolution of 0.01 gm
Graduated cylinders, 10, 25, 50 100 mL
Wire screen with holes smaller than 0.08 in (2 mm)
Stopwatch
Wire punch 1.0 mm (0.04 in) diameter by 50 mm–75 mm (2.5–3.0 in) length

Truncated cone centering guide with 1.5 mm (0.60 in) hole (plastic, brass, stainless steel)

Clean coarse sand passing No. 10 sieve (pea gravel may be substituted)

Graduated standpipe to measure hydraulic head

Compaction equipment; Harvard Miniature mold or similar cylinder with inside diameter of 1.3125 in (33.3 mm); outside diameter of 46.0 mm (1.8125 in); length of 71.5 m (2.816 in)

O-rings to fit end of test cylinder

Digital vernier calipers with a range of 150 mm (6 in) and a resolution of 0.1 mm (0.004 in)

Whatman #3 filter paper

Procedure

1. Compact or trim a soil specimen 38 mm (1.5 in) long into the pinhole test cylinder on top of the coarse sand and wire screen. If using a trimmed specimen, make sure that the specimen fits tightly into the test cylinder so no sidewall leakage will occur.
2. Insert the truncated cone centering guide into the top of the soil specimen using light finger pressure.

    Insert the 1 mm (0.039 in) wire punch into the centering guide and push the punch through the soil specimen. Make sure the punch passes completely through the soil specimen just into the coarse sand. The punch can be rotated while being pushed.
3. Remove the punch carefully.
4. Carefully place the wire screen on top of the specimen with the centering guide still in place and fill the remaining void in the top of the test cylinder with coarse sand.
5. Place O-rings on the ends of the test cylinder. Attach the top plate, connect the constant head source and place the assembly into the test fixture in a horizontal position.
6. Close the head source valve and open the valve connected to the head measuring standpipe.
7. Set the head at 2 in (50 mm) and open the valve on the head measuring standpipe and start the stopwatch.
8. Measure the effluent discharge flow using an appropriate graduated cylinder. Initially, the first two or three measurements of discharge should be made in units of time in seconds required to collect 10 mL of effluent. Alternatively, select a set time (such as 60 sec) and measure the amount of effluent collected during that time interval. (Note: if no flow occurs, disassemble the apparatus and re-punch the hole in the soil specimen.)
9. Subsequent measurements may consist of recording the time interval required to collect 25, 50 or 100 mL of effluent.

10. Observe the cloudiness of the effluent in the graduated cylinder from the side and from the top for each measured discharge. Record the cloudiness of the effluent as *very dark*, *dark*, *moderately dark*, *slightly dark*, *barely visible*, or *clear*.

11. Depending on the cloudiness of the effluent, the head may be increased to 180 mm (7 in), 380 mm (15 in) or 1020 mm (40 in) to modify the flow rate.

12. After the end of the test, dismantle the apparatus and extrude the soil specimen from the cylinder.

13. Cut or break the specimen open longitudinally and measure the average size of the hole using a digital caliper by comparing the hole against the needle used to punch the hole.

14. Classification of dispersive soils using this test method is based on the cloudiness of the effluent at the end of the test under different heads and final hole size. Table 17.1 gives ASTM recommendations for different classifications of Dispersion.

Calculations

There are no calculations with this test. The results are by visual observation.

Example

The results in Table 17.2 were obtained from a pinhole test from a sample of clay from Colorado.

Both tests would give a Dispersion classification of ND3 for this soil.

Discussion

The pinhole test is a direct performance test and is considered by some as the most reliable, despite the fact that the results are qualitative and somewhat subjective. The variations in head and flow are sufficiently broad in between the different classifications that there is little room for error in identifying a dispersive soil. In some cases, a 0.005N calcium sulfate $(CaSO_4)$ solution is used instead of distilled water to simulate groundwater.

A clear plastic test cylinder may be used instead of the Harvard miniature compaction mold. The cylinder should be machined to accept rubber O-rings on the ends and either fit into the end caps of the Harvard compaction equipment or into specially machined ends. Figure 17.2 shows a clear test chamber with machined plastic end caps. A schematic of the test setup is shown in Figure 17.3. A graduated cylinder to collect the effluent is placed on the right side of the test chamber after the valve.

Table 17.1 Criteria for Classifying Dispersive Soils (ASTM D4647)

Head (mm)	Test time (min.)	Final flow rate (mL/sec.)	Cloudiness from side	Cloudiness from top	Hole size after test (mm)	Description	Dispersion classification
50	5	1.0–1.4	Dark	Very dark	≥ 2.0	Highly dispersive	D1
50	10	1.0–1.4	Moderately dark	Dark	> 1.5	Dispersive	D2
50	10	0.8–1.0	Slightly dark	Moderately dark	≤ 1.5	Non-dispersive	ND4
180	5	1.4–2.7	Barely visible	Slightly dark	≥ 1.5	Slightly to moderately dispersive	ND3
380	5	1.8–3.2					
1020	5	> 3.0	Clear	Barely visible	< 1.5	Non-dispersive	ND2
1020	5	≤ 3.0	Perfectly clear	Perfectly clear	1.0	Non-dispersive	ND1

Table 17.2 Pinhole Test

Test	Head (mm)	Time (min)	Flow rate (ml/sec)	Cloudiness from side	Cloudiness from top	Final hole size (mm)
1	50	10	1.5	Barely visible	Slightly dark	≤1.5
2	180	5	1.7	Barely visible	Slightly dark	1.5

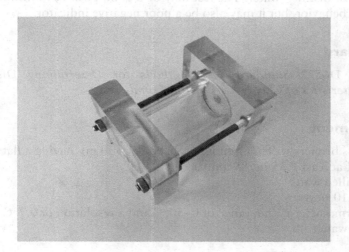

Figure 17.2 Plastic test chamber and end caps for performing pinhole test.

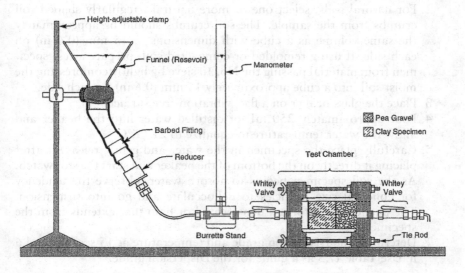

Figure 17.3 Schematic of test arrangement for performing the pinhole test.

DISPERSION BY CRUMB TEST

Introduction

The crumb test is the simplest of the tests used to identify dispersive soils but it also may be the least accurate. No special equipment is required and the test is purely qualitative, observing how a soil behaves when placed in a beaker of distilled water. The test may be a good positive indicator of dispersive behavior, but it may also be a poor negative indicator.

Standard

ASTM D6572 *Standard Test Methods for Determining Dispersive Characteristics of Clayey Soils by the Crumb Test*

Equipment

Glass beakers with a minimum capacity of 300 mL having a flat bottom of at least 3.35 in (85 mm) across
Distilled water
No. 10 sieve
Thermometer with a range of 0–50°C and a resolution of 0.2°C
Stopwatch

Procedure

1. Select a representative soil sample for testing.
2. For natural soils, select one or more natural irregularly shaped soil crumbs from the sample. The soil crumbs should be approximately the same volume as a cube with dimensions of 15 mm (0.6 in) on each side. If using remolded or compacted soil, prepare a test specimen from material passing the No. 10 sieve by lightly compressing the moist soil into a cube approximately 15 mm (0.6 in) on each side.
3. Place the glass beaker on a flat, vibration-free surface.
4. Pour approximately 250 mL of distilled water into the beaker and allow the water temperature to equalize to $21 \pm 6°C$.
5. Carefully place the specimen in the water and gently release it after placing it directly on the bottom of the beaker and start the stopwatch.
6. As the soil specimen begins to absorb water, observe the tendency for colloidal size particles to deflocculate and go into suspension. The particles will form a dense cloud or halo that extends from the specimen.
7. Determine the Dispersion grade and temperature at 15 sec, 1 h and 6 h. Use Table 17.3 as a guide for the Dispersion grade.

Table 17.3 Categories of Dispersion by Observations in Crumb Test

Dispersion category	Characteristics
Grade 1 Non-dispersive	No reaction; soil may crumble, slake and spread out, but there is no turbid water created by colloids suspended in water. All particles settle within the first hour.
Grade 2 Intermediate	Slight reaction; a faint barely visible colloidal suspension causes turbid water near portions of the soil crumb surface. If the cloud is easily visible, assign Grade 3. If the cloud is faintly seen in only one small area, assign Grade 1.
Grade 3 Dispersive	Moderate reaction; an easily visible cloud of suspended colloids is seen around the outside of the soil crumb surface. The cloud may extend up to 0.4 in (10 mm) away from the soil crumb mass along the bottom of the beaker.
Grade 4 Highly dispersive	Strong reaction; a dense profuse cloud of suspended colloidal particles is seen around the entire bottom of the beaker. Occasionally the soil crumb Dispersion is so extensive that it is difficult to determine the interface of the original crumb and the colloidal suspension. Often the colloidal suspension is easily visible on the sides of the beaker.

Note: If the Dispersion grade changes during the tests, the 1 h observation is normally used for the overall test evaluation. However, if the grade changes from 2 to 3 or from 3 to 4 between the 1 h and 6 h observations, use the 6 h observation.

Calculations

There are no calculations with this test. The results are obtained by visual inspection.

Discussion

Soils that have a Grade 1 or 2 Dispersion category occasionally are shown to be dispersive in other tests or in field performance. The results of the crumb test can be used as a quick screening test to determine if either the double hydrometer or pinhole test should be performed to provide further evaluation of dispersivity.

REFERENCES

Bell, F.G., and Maud, R.R., 1994. Dispersive Soils: A Review from a South African Perspective. *Quarterly Journal of Engineering Geology*, Vol. 27, pp. 195–210.
Decker, R.S., and Dunnigan, L.P., 1977. Development and Use of the Soil Conservation Service Dispersion Test. *ASTM STP*, Vol. 623, pp. 94–109.

Elges, H.F., 1985. Problem Soils in South Africa: State of the Art. *The Civil Engineer in South Africa*, Vol. 27, No. 7, pp. 347–535.

Lefabvre, G., Rohan, K., and Douville, S., 1985. Erosivity of Natural and Intact Structured Clay: Evaluation. *Canadian Geotechnical Journal*, Vol. 22, pp. 508–517.

Sherard, J.L., Dunnigan, L.P., Decker, R.S., and Steele, E.F., 1976. Identification and Nature of Dispersive Soils. *Journal of the Geotechnical Engineering Division, ASCE*, Vol. 102, No. 4, pp. 287–301.

Sherard, J.L., Dunnigan, L.P., Decker, R.S., and Steele, E.F., 1976a. Pinhole Test for Identifying Dispersive Soils. *Journal of the Geotechnical Engineering Division, ASCE*, Vol. 102, No. 1, pp. 69–85.

ADDITIONAL READING

Dascal, O., Pouliot, G., and Hurtubise, J., 1977. Erodibility Tests on a Sensitive Cemented Marine Clay (Champlain Clay). *ASTM STP*, Vol. 623, pp. 74–93.

Elges, H.F., 1985. Dispersive Sols. *The Civil Engineer in South Africa*, Vol. 29, pp. 397–399.

Forsythe, P., 1977. Experiences in Identification and Treatment of Dispersive Clays in Mississippi. *ASTM STP*, Vol. 623, pp. 135–155.

Goldsmith, P.R., and Smith, E.H., 1984. Tunnelling Soils in South Auckland, New Zealand. *Proceedings of the 4th Australia-New Zealand Conference on Geomechanics*, pp. 445–449.

Heinzen, R.T., and Arulanandan, K., 1977. Factors Influencing Dispersive Clays and Methods of Identification. *ASTM STP*, Vol. 623, pp. 202–217.

Lewis, D.A., and Schmidt, N.O., 1977. Erosion of Unsaturated Clay in a Pinhole Test. *ASTM STP*, Vol. 623, pp. 260–273.

Maharaj, A., and Paige-Green, P., 2013. The SCS Double Hydrometer Test in Dispersive Soil Identification. *Proceedings of the 18th International Conference on Soil Mechanics and Geotechnical Engineering*, pp. 389–392.

McDonald, L.A., Stone, P.C., and Ingles, O.G., 1981. Practical Treatment for Dams in Dispersive Soil. *Proceedings of the 10th International Conference on Soil Mechanics and Foundation Engineering*, Vol. 2, pp. 355–360.

Riley, P.B., 1977. Dispersion in Some New Zealand Clays. *ASTM STP*, Vol. 623, pp. 354–361.

Sherard, J.L., Decker, R.S., and Ryker, N.L., 1972. Piping in Earth Dams of Dispersive Clay. In *Performance of Earth and Earth Supported Structures, ASCE*, pp. 589–626.

Chapter 18

Initial Consumption of Lime

BACKGROUND

The addition of lime to fine-grained soils to improve their engineering properties and behavior has been a common method of ground improvement since the 1930s. The reaction of soil + lime is complex and depends on the mineralogy, Cation Exchange Capacity, Specific Surface Area, clay content and other soil constituents, such as Carbonate Content. The addition of lime to soil also results in an increase in pH of the mixture to a maximum value of 12.4 (lime–water mixture). At this pH the solubility of silica and alumina are significantly increased.

The reactions are generally considered in two areas: 1) modification or alteration (short-term), and 2) stabilization or solidification (long-term). The improvement of the engineering properties is the result of four types of reactions: 1) exothermic reaction; 2) cation exchange; 3) flocculation and agglomeration; and 4) long-term pozzolanic (cementing) reactions. The initial reaction between quicklime and soil water produces an exothermic reaction that produces heat:

$$CaO + H_2O \text{ gives } Ca(OH)_2 + Heat \tag{18.1}$$

The reaction consumes water from the soil matrix, and the heat has a drying action that causes some immediate improvement in workability.

The mechanisms of lime reaction with soil depend on the type of lime used and the soil. The most frequently used limes in stabilization are quicklime (CaO) and hydrated lime ($Ca(OH)_2$). Quicklime for use in soil stabilization is produced by heating limestone ($CaCO_3$) to remove carbon dioxide (CO_2), leaving calcium oxide (CaO); hydrated lime is produced by running crushed quicklime through a water hydrator to form calcium hydroxide ($Ca(OH)_2$). Agricultural lime (calcium carbonate $CaCO_3$) will not react with soils and should not be used in stabilization of fine-grained soils.

Eades and Grim (1966) suggested a simple indirect laboratory test method for establishing the optimum lime content. The test involves measuring the pH of soil–lime mixtures at various lime content and determining the point

DOI: 10.1201/9781003263289-20

at which no further increase in pH occurs. This technique actually gives an upper bound value of the lime addition, but other more direct tests, such as plasticity or strength tests, may show lower amounts of lime needed, depending on the specific design requirements of a project.

INTRODUCTION

The Initial Consumption of Lime (ICL) test measures the pH of a soil + lime suspension using different percentages of lime added to the soil. Either hydrated lime ($CaOH_2$) or quicklime (CaO) may be used for the test. A saturated solution of lime in distilled water completely free of carbon dioxide has a pH value of 12.4 at 25°C. This pH is required to maintain reaction between lime and any reactive components in the soil. Different percentages of lime by air-dry mass of the soil are mixed with water, and the minimum amount of lime needed to give a pH of 12.4 is defined as the ICL of the soil. This value is used as an initial guide to the amount of lime needed in the field to be mixed with the soil.

STANDARD

ASTM D6272 *Standard Test Method for Using pH to Estimate the Soil-Lime Proportion Requirement for Soil Stabilization*, BS 1924-2

EQUIPMENT

pH meter
Electronic scale with a minimum capacity of 600 gm and a resolution of
   0.01 gm
Glass bottles with lids with a capacity of 250 mL
Distilled water with pH = 7
Laboratory grade lime ($CaOH_2$ or CaO)
100 mL graduated cylinder
Standard pH reference solution with pH = 12.0
Glass stirring rod
Ceramic mortar and rubber-tipped pestle
Aluminum tares or small ceramic bowls

PROCEDURE

1. Calibrate the pH meter according to the manufacturer's instruction using a standard pH solution of 12.0.

2. Prepare a saturated solution of lime by placing 5 gm of lime in a clean glass bottle. Add 100 mL of distilled water and stir or shake for 30 sec every 10 min for 1 h. After 1 h, determine the pH. (Note: Always add water to dry lime; never add dry lime to water.)
3. Select a representative sample of soil and allow the sample to air dry.
4. After air drying, crush the sample using a mortar and rubber-tipped pestle to pass the No. 40 sieve.
5. Take the dried and sieved soil and weigh out test specimens of 20 gm to the nearest 0.01 gm for as many different lime percentages as there are to be tested.
6. Place each test specimen into a clean glass bottle. Weigh out lime to the nearest 0.01 gm according to the percentage to be used in each test bottle; e.g., 1% = 0.20 gm by dry soil mass; 2% = 0.4 gm, etc. Usually, lime contents of 1%, 2%, 3%, 4%, 6% and 8% are sufficient.
7. Add the lime to each bottle and stir to mix the soil with the lime.
8. Add 100 mL of distilled water to each bottle. Stir the bottles until there is no evidence of dry soil on the bottom of the bottle. Place the lid tightly on the top of each bottle. Shake the bottles for 30 sec every 10 min for 1 h.
9. After 1 h, determine the pH of the suspensions.
10. After testing, dispose of the mixtures in an appropriate container.

CALCULATIONS

There are no calculations for this test. The pH meter is read directly.

EXAMPLE

ICL tests were performed on plastic Buckshot Clay and gave the results in Table 18.1. A plot of pH vs. % lime is shown in Figure 18.1.

Table 18.1 ICL Tests

% CaO	pH
0	7.64
1	11.79
2	12.25
3	12.36
4	12.40
6	12.43
8	12.44
10	12.45

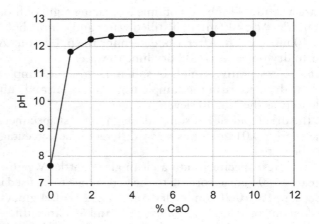

Figure 18.1 Plot of pH vs. lime content.

For this soil, the ICL is estimated to be approximately 4%.

DISCUSSION

Air-dried soil is used in this test instead of oven-dried soil as a matter of practicality and keeping with the precision of the test procedure. (The difference between using air-dried and oven-dried soil will not affect the outcome of the test.) ASTM recommends using 25 gm of soil for the test, although the author has found no significant difference when using 20 gm. The use of 20 gm of soil is also keeping in line with the British Standard BS1924-2:1990 *Determination of Initial Consumption of Lime*. The ICL test is an approximate method and is intended to give initial guidance in selecting the minimum amount of lime to be added to soil to produce beneficial reactions and to identify if other more detailed tests may need to be performed to determine the optimum lime content needed for a specific project. ASTM D6276 notes that in some soils the pH may not go above 12.3 and may be the result of the soil holding univalent cations, such as sodium.

There are some subtle differences in the test procedure originally suggested by Eades and Grim (1960) and ASTM D6276 and BS 1924-2. Table 18.2 gives a comparison of the different methods.

Table 18.2 Comparison between Different Test Methods

	Eades and Grim (1960)	ASTM D6276	BS 1924-2
Soil size fraction	<No. 40	<No. 40	<No. 40
Mass of soil	20.0 gm	25.0 gm	20.0 gm
Volume of water	100 mL	100 mL	100 mL
Oven dried?	Oven dried equivalent	Air dried (May be oven dried)	Oven dried
Bottles	Plastic (≥150 mL)	Plastic (≥150 mL)	Glass or plastic (250 mL)
Time	1 h	1 h	1 h
pH buffer standard	12.0	12.0	9.2

REFERENCE

Eades, J., and Grim, R., 1966. A Quick Test to Determine Lime Requirements for Lime Stabilization. *Highway Research Record*, Vol. 139, pp. 61–73.

ADDITIONAL READING

Cambi, C., Guidobaldi, G., Cecconi, M., Comodi, P., and Russo, G., 2016. On the ICL Test in Soil Stabilization. *Proceedings of the 1st International Workshop for Metrology for Geotechnics*, pp. 31–34.
Clauss, K.A., and Loudon, P.A., 1971. The Influence of Initial Consumption of Lime and the Stabilization of South African Road Materials. CSIR Report National Institute for Road Research.

Chapter 19

Lime Fixation Point

INTRODUCTION

For predominantly fine-grained soils, the Lime Fixation Point (LFP) is often used to determine the economic optimum lime content to be added to a soil to provide improvement. The Lime Fixation Point is based on changes in the plasticity of the soil as determined by the Atterberg Limits and is defined as the amount of lime that produces no further change in Plastic Limit (Hilt & Davidson 1960). The addition of lime up to the LFP helps with immediate modification, while lime in excess of the LFP is available for long-term pozzolanic reactions. Calcium of the lime reacts with the soil to form a gel of calcium silicate hydrate (CSH) and calcium alumina hydrate (CAH) that cements the soil particles together. The procedure is simple and provides tests quickly at low cost. It is also based on directly evaluating the change in soil behavior (plasticity) produced by the addition and mixing of lime.

STANDARD

There is no ASTM standard test method for this test. The test procedure described in this chapter generally follows the method used by Hilt and Davidson (1960), except that the Plasticity Index is determined immediately after the addition of lime to the soil. No time is allowed for long-term lime-clay reactions. As the thread method for determining Plastic Limit was used in these tests by Hilt and Davidson (1960), this procedure is described in this chapter.

EQUIPMENT

Ceramic mixing bowls
Clean glass plate for rolling Plastic Limit
Stainless steel reference rod of 3.2 mm (1/8 in) diameter
Casagrande Liquid Limit cup

DOI: 10.1201/9781003263289-21

Electronic scale with a minimum capacity of 600 gm and a resolution of 0.01 gm

Ceramic or aluminum moisture tares

Spatula

Plastic wrap

Laboratory grade lime ($CaOH_2$ or CaO)

Convection oven capable of maintaining $110° \pm 5°C$

PROCEDURE

1. Select a representative sample of soil. Separate the soil into six equal parts with sufficient mass to perform the Plastic Limit test using the thread method described in Chapter 12.
2. Determine the Water Content of the soil using either the oven dry method or microwave oven method described in Chapter 1.
3. Place each of the specimens in a ceramic bowl and cover with plastic wrap to prevent moisture loss.
4. Add sufficient distilled water to each specimen to bring the consistency close to the Liquid Limit. This can be checked with the Casagrande Liquid Limit cup. Cover and allow the soil–water mixtures to temper for 24 hrs.
5. After 24 hrs. add different percentages of lime by dry mass to each of the six specimens; usually, 0%, 1%, 2%, 3%, 4%, 6%, 8% and 10% lime is sufficient. Mix the lime into the soil thoroughly with a spatula until no more lime can be seen in the mixture. (Note: For more detailed delineation of the LFP, smaller amounts of lime may be used, e.g., 0%, 0.5%, 1%, 1.5%, 3%, 3%, 4%, 5%).
6. Determine the Plastic Limit of each specimen using the thread method described in Chapter 12.
7. After testing, dispose of any excess soil in an appropriate container.

CALCULATIONS

Calculate the Plastic Limit (Water Content) of each specimen using the mass of wet soil and dry soil for each Plastic Limit determination as described in Chapter 1.

EXAMPLE

The Plastic Limit of a sample of London Clay was determined with different percentages of CaO. Results of the tests are shown in Table 19.1. The graph below shows a plot of PL vs. % Lime in Figure 19.1.

Table 19.1 Plastic Limit for Different Lime Content

% CaO	PL
0	37
1	45
2	51
3	53
4	55
6	55
8	56

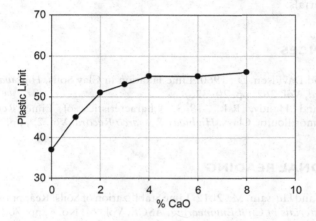

Figure 19.1 Change in Plastic Limit with addition of lime (CaO).

The results indicate that the Lime Fixation Point is approximately 4% for this soil.

DISCUSSION

The test procedure described is a very practical approach at estimating the economic lime content. Results may be slightly different if hydrated lime is used instead of quicklime. The results are approximate and only consider changes in plasticity. If other soil characteristics are of interest for a particular project, such as shear strength, the results of this test can be used as a guide in preparing other kinds of specimens. The results show that an immediate improvement in field workability can be achieved with just a very small amount of lime. This is lime modification.

In most soils, the lime does not substantially change the Liquid Limit, but this means that the Plasticity Index is decreased because of the increase in Plastic Limit. In some cases, the soil may actually become non-plastic as a result of the lime addition. Ho and Handy (1963) referred to the optimum lime content where the Plastic Limit stops changing as the Lime Retention Point.

A simple modification to the test procedure that can be used to evaluate long-term behavior would be to cover the soils after mixing with lime and allow then to set for 7 days and then perform the Plastic Limit test. These results may show that less lime is required to produce soil stabilization effects.

Bare hands should not be used to roll out the soil as the lime may cause irritation to the skin. Lime is caustic and care should be used in handling any soil treated with lime or lime itself. Do not rub your eyes after handling these materials.

REFERENCES

Hilt, G., and Davidson, D., 1960. Lime Fixation in Clay Soils. *Highway Research Record*, Vol. 262, pp. 20–32.

Ho, C., and Handy, R.L., 1963. Characteristics of Lime Retention by Montmorillonitic Clays. *Highway Research Record*, Vol. 29, pp. 55–69.

ADDITIONAL READING

Dash, S.K., and Hussain, M., 2012. Lime Stabilization of Soils: Reappraisal. *Journal of Materials in Civil Engineering, ASCE*, Vol. 24, No. 6, pp. 707–714.

Hussain, M., and Dash, S.K., 2010. Influence of Lime on Plasticity Behaviour of Soils. *Proceedings of the Indian Geotechnical Conference*, pp. 537–540.

Sivapuliaiah, P.V., Prashanth, J.P., and Sridharan, A., 1988. Delay in Compaction and Importance of Lime Fixation Point on the Strength and Compaction Characteristics of Soil. *Proceedings of the Institution of Civil Engineers, Ground Improvement*, Vol. 2, No. 1, pp. 27–32.

Chapter 20

Hygroscopic Water Content

INTRODUCTION

Hygroscopic Water Content is a behavioral characteristic of soils that essentially describes (in quantitative terms) how much water can be adsorbed on the internal and external surface area of clay minerals present in the soil. The Hygroscopic Water Content of a soil is the same as the air-dried Water Content (i.e., Water Content when exposed to the atmosphere for a long period of time at constant relative humidity). However, since atmospheric Water Content changes with changes in the weather, the Hygroscopic Water Content of soils is not constant but varies with humidity. Measurement of the Hygroscopic Water Content was in practice at least as early as the 1930s according to Pichler (1953).

A number of studies have shown that the Hygroscopic Water Content depends primarily on the amount and type of clay minerals present in the soil. This is fundamentally related to the Specific Surface Area (SSA) (i.e., total surface area) and Cation Exchange Capacity (CEC). The test is simple, is easy to perform, and requires only basic laboratory equipment.

STANDARD

There is currently no ASTM standard test method for this test. The test procedure described in this chapter is a modification of the procedures discussed by Shah & Singh (2005) and Prakash et al. (2016).

EQUIPMENT

Large glass desiccator
Electronic scale with a minimum capacity of 250 gm and a resolution of 0.01 gm
Ceramic or aluminum moisture tares
Hygrometer that is capable of reading relative humidity to nearest 1%

DOI: 10.1201/9781003263289-22

Thermometer with a range of 10 to 50°C and a resolution of 0.5°C

PREPARATION OF HUMIDITY CHAMBER

Specimens of oven-dried soil must be placed in a humid chamber that maintains constant humidity to allow the soil to adsorb water. Several different arrangements may be used. A specially fabricated humidity chamber in which the humidity can be regulated may be used. Alternatively, a glass desiccator may also be used and is the simplest. A saturated solution of NaCl (36 gm NaCl/100 mL water) placed in the bottom of a desiccator produces a relative humidity of 75%. The bottom of the desiccator may also be filled with distilled water to about 1 in. (25.4 mm) below the desiccator tray.

PROCEDURE

1. Select a representative sample of air dried soil passing the No. 40 sieve.
2. Place the soil in a small ceramic bowl and place in a convection oven at 110°C for 24 h.
3. After oven drying, remove about 10–15 gm of soil and place it in a small moisture tare of known mass. (Note: Smaller amounts of soil (4–5 gm) may be used to reduce testing time.)
4. Determine the total mass of the oven dry soil + tare to the nearest 0.01 gm.
5. Place the soil and moisture tare with the lid off on the tray in the desiccator. Place the lid tightly on the desiccator.
6. Periodically, remove the moisture tare from the desiccator and determine the total mass to the nearest 0.01 gm.
7. Continue taking measurements of total mass reaches equilibrium, i.e., until the different between consecutive measurements varies less than 0.05 gm for a period of at least 2 h. (Note: Some soils will reach equilibrium mass after a few hours, while others may take several days.)
8. Record the final total mass to the nearest 0.01 gm.
9. Record the relative humidity.

CALCULATIONS

Calculate the Hygroscopic Water Content as:

$$W_H = (M_{WET} - M_{DRY})/(M_{DRY} - M_{TARE}) \times 100\% \qquad (20.1)$$

Report the value of W_H to the nearest 0.1%

EXAMPLE

A sample of Black Cotton clay from India was tested at a relative humidity of 35%. The results are given in Table 20.1.

The initial mass of the oven dry soil + tare = 32.12 gm
The mass of the tare = 1.29 gm
The dry mass of soil = 30.83 gm

A plot of the test results are shown in Figure 20.1. $W_H = 5.6\%$ at 35% RH.

DISCUSSION

As previously noted, the testing time can be reduced considerably by using a smaller specimen. However, using a smaller specimen may reduce the accuracy of the measurements as a smaller specimen will adsorb less water.

The test described in this chapter is easy to perform, however the results depend on the relative humidity. Tests should be performed at constant temperature between 22°C and 25°C. Figure 20.2 shows an example of the influence of relative humidity on two different clays, a kaolinite and a montmorillonite. The relationship is non-linear and is basically a vapor isotherm. This means that it is important that the relative humidity be maintained

Table 20.1 Sample of Black Cotton Clay

Reading no.	Time (min.)	Total mass (gm)	Mass of water (gm)	Dry mass (gm)	Water Content (%)
1	0	32.12	0.00	30.83	0.0
2	38	32.32	0.10	30.83	0.3
3	60	32.39	0.27	30.83	0.9
4	120	32.56	0.34	30.83	1.4
5	240	32.81	0.69	30.83	2.2
6	360	33.01	0.89	30.83	2.9
7	480	33.18	1.06	30.83	3.4
8	720	33.42	1.30	30.83	4.2
9	1440	33.70	1.58	30.83	5.1
10	1800	33.73	1.61	30.83	5.2
11	2209	33.78	1.66	30.83	5.4
12	2880	33.81	1.69	30.83	5.5
13	3660	33.85	1.73	30.83	5.6
14	4421	33.86	1.74	30.83	5.6
15	5121	33.85	1.73	30.83	5.6
16	7351	33.86	1.73	30.83	5.6

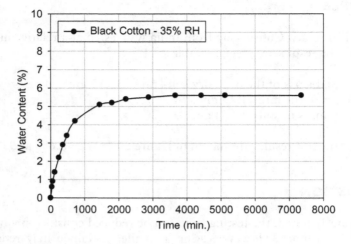

Figure 20.1 Hygroscopic Water Content vs. time.

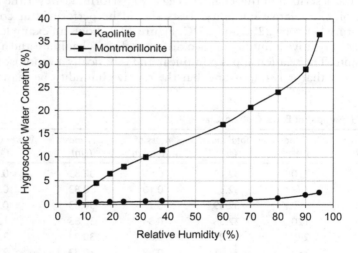

Figure 20.2 Influence of relative humidity on the Hygroscopic Water Content of two clays (data from Barshad 1955.)

constant throughout the duration of the test. Data from the literature indicate that tests for Hygroscopic Water Content have been performed with relative humidity in the range of 20%–100%.

Several studies have shown strong correlations between Hygroscopic Water Content and other soil characteristics e.g., Davidson and Sheeler 1952; Barshad 1955; Robinson et al. 2002; Arthur et al. 2021). Figure 20.3

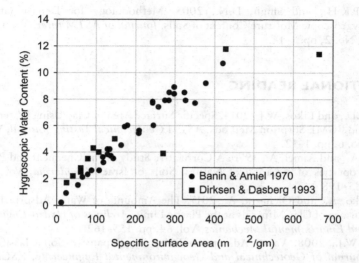

Figure 20.3 Correlation between SSA and Hygroscopic Water Content.

shows an example of a correlation reported between Specific Surface Area (SSA) to Hygroscopic Water Content.

REFERENCES

Arthur, E., Ur Rehman, H., Tuller, M., Pouladi, N., Norgaard, T., Moldrup, P., and de Jonge, L.W., 2021. Estimating Atterberg Limits of Soils from Hygroscopic Water Content. *Geoderma*, Vol. 381, p. 9.

Barshad, I., 1955. Adsorptive and Swelling Properties of Clay Water Systems. *Proceedings of the 1st National Conference on Clays and Clay Technology*, pp. 70–75.

Davidson, D.T., and Sheeler, J.B., 1952. Clay Fraction in Engineering Soils: Influence of Amount on Properties. *Highway Research Board Proceedings*, Vol. 31, pp. 558–563.

Pichler, E., 1953. The Expansion of Soils Due to the Presence of Clay Minerals as Determined by the Adsorption Test. *Proceedings of the 3rd International Conference on Soil Mechanics and Foundation Engineering*, Vol. 1, pp. 43–46.

Prakash, K., Sridharan, A., and Sudheendra, S., 2016. Hygroscopic Moisture Content: Determination and Correlations. *Environmental Geotechnics, Proceedings of the Institute of Civil Engineers, ICE*, Vol. 3, No. 5, pp. 293–301.

Robinson, D.A., Cooper, J.D., and Gardner, C.M., 2002. Modelling the Relative Permittivity of Soils Using Soil Hygroscopic Water Content. *Journal of Hydrology*, Vol. 255, pp. 39–49.

Shah, P.K.H., and Singh, D.N., 2005. Methodology for Determination of Hygroscopic Moisture Content of Soils. *Journal of ASTM International*, Vol. 3, No. 2, pp. 1–14.

ADDITIONAL READING

Akin, I.D., and Likos, W.J., 2014. Specific Surface Area of Clay Using Water Vapor and EGME Sorption Methods. *ASTM Geotechnical Testing Journal*, Vol. 37, No. 6, pp. 1–12.
Banin, A., and Amiel, A., 1970. A Correlative Study of the Chemical and Physical Properties of a Gropu of Natural Soils of Israel. *Geoderma*, Vol. 3, pp. 185–198.
Goraczko, A., and Olchawa, A., 2017. The Amounts of Water Adsorbed to the Surface of Clay Minerals at the Plastic Limit. *Archives of Hydro-Engineering and Environmental Mechanics*, Vol. 64, pp. 155–162.
Likos, W.J., 2008. Vapor Adsorption Index for Expansive Soil Classification. *Journal of Geotechnical and Geoenvironmental Engineering, ASCE*, Vol. 134, No. 7, pp. 1005–1009.
Moiseev, K.G., 2008. Determination of Specific Surface Area from the Hygroscopic Water Content. *Eurasian Soil Science*, Vol. 41, No. 7, pp. 744–748.
Newman, A.C., 1983. The Specific Surface Area of Soils Determined by Water Sorption. *Journal of Soil Science*, Vol. 34, pp. 23–32.

Chapter 21

Soil Suction Using Filter Paper

INTRODUCTION

Natural soils above the water table and most compacted soils are unsaturated; that is, some of the void space is filled with air. These soils have an affinity for additional water and have internal suction or negative pore water pressure. Soil matric suction is important in the interpretation of unsaturated soil properties and behavior. In natural soil profiles, the soil above the water table develops negative pore water pressure or suction. Figure 21.1 shows measured soil suction at a varved clay site in Amherst, Ma. during two different seasons, fall and spring. Soil suction was measured using in place tensiometers. The water table is characteristically high during the spring and low in the late summer. The pore pressure profiles reflect the seasonal change in pore water pressure and soil suction at these times.

The laboratory test method described in this chapter provides a simple procedure to measure the amount of the soil matric suction using filter papers as passive sensors. The test procedure may be used to determine soil matric suction values over the range of about 80–6000 kPa. The method controls the variables for measurement of Water Content of filter paper that is in direct contact with soil. The filter paper is enclosed with a soil specimen in an airtight container until moisture equilibrium from water vapor in the container is reached. The test uses very simple equipment and is inexpensive, but several days are usually required for the equilibrium to take place. Tests may be performed on undisturbed specimens trimmed from undisturbed tube or block samples and may also be performed on compacted samples.

STANDARD

ASTM D5298 *Standard Test Method for Measurement of Soil Potential (Suction) Using Filter Paper*

DOI: 10.1201/9781003263289-23

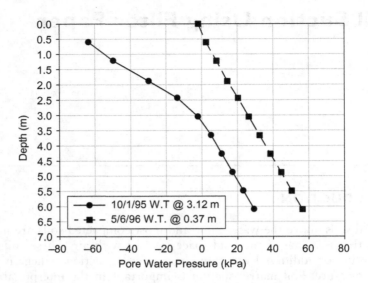

Figure 21.1 Pore water pressure profiles in Amherst, Ma.; fall and spring.

EQUIPMENT

Filter paper (Whatman No. 42; Fisherbrand 9-790A; Schleicher and Schuell No. 589 White Ribbon)

Glass or aluminum specimen container 120 to 240 mL capacity with lid

Filter paper container

Insulated chest of about 1 ft³ (0.03 m³) capacity (polystyrene)

Electronic analytical scale with a minimum capacity of 20 gm and a resolution of 0.0001 gm

Convection oven capable of maintaining 110° ± 5°C

Electrical tape

Tweezers

Metal block with a flat surface and a mass > 500 gm

Thermometer with a range of 0 to 50°C and a resolution of 0.5°C

Spatula or trimming knife

Desiccator with silica gel or calcium sulfate desiccant (Drierite)

FILTER PAPER CALIBRATION

Filter papers used in this test method should be calibrated to establish the relationship between filter paper Water Content and soil suction. ASTM D5298 provides a detailed procedure for developing the calibration curve for filter papers to be used. Alternatively, ASTM D5298 provides calibration curves for Whatman No. 42 and Schleicher and Schuell No. 589 filter paper. These curves can be used with reasonable accuracy.

Note that the calibration curves shown in Figure 21.2 have two straight line segments; one for w < about 45% (Whatman 42) and another section above w = 45%. Similarly, the calibration for Schleicher and Schuell No. 589 changes at about W = 50%. Other calibrations are available (see discussion).

PROCEDURE

1. Dry filter papers that will be used for measurement in this test for 16 h or overnight in a convection oven at 110°C.
2. Select a soil specimen of about 200–400 gm for testing. Trim the ends of the specimen with a spatula or trimming knife. (Note: Specimens are typically trimmed from undisturbed tube samples or from compacted samples extruded from a compaction mold or obtained in the field.)
3. Place three stacked filter papers in contact with two sections of the specimen as shown in Figure 21.3. The center paper is used for analysis of matric suction. The outer papers prevent soil contamination of the center paper. The inner paper should be slightly smaller in diameter (0.08–0.12 in [2–3 mm]) than the outer papers.
4. Place the lid onto the top of the specimen container and seal tightly with at least two wraps of plastic electrical tape.

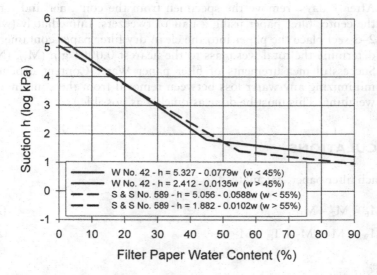

Figure 21.2 ASTM D5298 Calibration curves of suction-Water Content for wetting of filter paper (after Greacen et al. 1987)

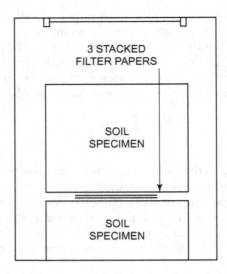

Figure 21.3 Placement of stacked filter papers on specimens.

5. Placed the sealed container in an insulated chest and place the chest in a location with a nominal constant temperature of 20°C.
6. Allow the filter paper and the specimen to come to equilibrium for 7 days.
7. Select a clean, dry cold filter paper container and determine the mass to the nearest 0.0001 gm (M_C).
8. After 7 days, remove the specimen from the container and remove the center filter paper using a pair of tweezers. Immediately (within 2–3 sec) place the paper into the clean dry filter paper container and determine the total wet mass to the nearest 0.0001 gm (M_1). (Note: Successful measurements of filter paper Water Content depend on minimizing any water loss between removal from the container and weighing. This must be done as quickly as possible.)

CALCULATIONS

For each filter paper, calculate:

$$M_F = M_2 - M_H \tag{21.1}$$

$$M_W = M_1 - M_2 + T_H - M_C \tag{21.2}$$

where:

M_F = mass of dry filter paper
M_1 = total mass of container with wet filter paper

M_2 = total dry mass
M_H = mass of the hot container
M_W = mass of water in the filter paper
M_C = mass of the cold container

Calculate the Water Content of the filter paper as:

$$W_F = \left[(M_W/M_F)\right] \times 100\% \tag{21.3}$$

where:
   W_F = Water Content of the filter paper

Calculate the suction from the calibration curve and report the suction to the nearest whole number.

EXAMPLE

A sample of glacial till from Ames, Iowa, was collected using a 3 in (76 mm) thin wall Shelby tube. After extruding the sample, a specimen was trimmed and tested to determine the soil suction. A Whatman No. 42 filter paper with a diameter of 2.0 in (51 mm) was used. The sample was allowed to equilibrate for 7 days.

   Dry mass of Whatman No. 42 = 0.2231 gm
   Wet mass of Whatman No. 42 = 0.2872 gm
   W_F = [(0.2872 gm – 0.2231gm)/ 0.2231 gm] × 100% = 28.7%
   Using the calibration chart given in ASTM D4298, S = 5.327 – 0.0779(28.7) = $10^{3.09}$ = –1230 kPa
   Using the calibration equation from Kim et al. (see discussion) S = 5.336 – 0.779(28.7) = $10^{3.10}$ = –1259 kPa
   A negative sign is used to indicate that the pressure is a suction.

DISCUSSION

The quality of matric suction measurements using the filter paper method depends on the quality of the measurements of the wet mass of the filter paper following moisture equilibrium. Likewise, this depends on providing a good seal on the specimen container during the equilibrium period.

   Other calibration curves are available for converting filter paper Water Content to matric suction. Chandler and Gutierrez (1986) suggested a simple linear expression for Whatman No. 42 filter paper as:

$$pF = 5.850 \quad 0.06622W \tag{21.4}$$

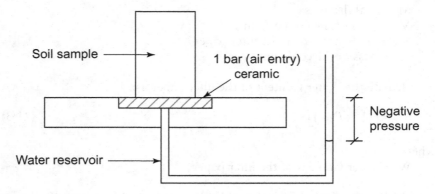

Figure 21.4 Fixture used to measure soil suction (from Ridley & Burland 1993).

where:
  pF = log soil suction (kPa)
  W = Water Content

Marinho and Oliveira (2006) suggested:

$$pF = 4.83 \quad 0.0839W \quad (\text{for } W < 33\%) \qquad (21.5a)$$

$$pF = 2.57 \quad 0.0154W \quad (\text{for } W > 33\%) \qquad (21.5b)$$

A summary of a number of calibration equations has been presented by Power et al. (2008). The most recent calibration for Whatman No. 42 filter paper has been presented by Kim et al. (2017) as:

$$\log_{10}(S) = 5.336 \quad 0.0779W \quad (\text{for } W < 45.5\%) \qquad (21.6a)$$

$$\log_{10}(S) = 2.394 \quad 0.0132W \quad (\text{for } W > 45.5\%) \qquad (21.6b)$$

where:
  S = suction (kPa)

An alternative to the filter paper method is to use a pressure transducer to directly measure the soil suction, as shown in Figure 21.4. Test methods using this approach have been suggested by Ridley and Burland (1993; 1999) and appear to be reliable for suction values as high as 1500 kPa (215 psi). The equipment is easy to fabricate and simple to use for a wide range of applications (Ridley et al. 2003). This technique has also been used to evaluate sample disturbance (e.g., Poirier et al. 2005).

REFERENCES

Chandler, R.J., and Gutierrez, C.I., 1986. The Filter Paper Method of Suction Measurement. *Geotechnique*, Vol. 36, No. 2, pp. 265–268.

Greacen, E.L., Walker, G.R., and Cook, P.G., 1987. Procedure for the Filter Paper Method of Measuring Soil Suction. *Proceedings of the International Conference on Measurement of Soil and Plant Water Status*, pp. 137–143.

Kim, H., Prezzi, M., and Saldago, R., 2017. Calibration of Whatman Grade 42 Filter Paper for Soil Suction Measurement. *Canadian Journal of Soil Science*, Vol. 97, pp. 93–97.

Marinho, F.A., and Oliveira, O.M., 2006. The Filter Paper Method Revised. *Geotechnical Testing Journal, ASTM*, Vol. 29, No. 3, pp. 250–258.

Poirier, S.E., DeGroot, D.J., and Sheahan, T.C., 2005. Measurement of Suction in a Marine Clay as an Indicator of Sample Disturbance. *Proceedings of GeoFrontiers Congress 2005, ASCE*, pp. 1–10.

Power, K.C., Vanapalli, S.K., and Garga, V.K., 2008. A Revised Contact Filter Paper Method. *Geotechnical Testing Journal, ASTM*, Vol. 31, No. 6, pp.461–469.

Ridley, A.M., and Burland, J.B., 1993. A New Instrument for the Measurement of Soil Suction. *Geotechnique*, Vol. 43, pp. 321–324.

Ridley, A.M., and Burland, J.B., 1999. Discussion of Use of the Tensile Strength of Water for Direct Measurement of High Soil Suction. *Canadian Geotechnical Journal*, Vol. 36, pp. 178–180.

Ridley, A.M., Dineen, K., Burland, J.B., and Vaughn, P.R., 2003. Soil Matrix Suction: Some Examples of its Measurement and Application in Geotechnical Engineering. *Geotechnique*, Vol. 53, No. 2, pp. 241–254.

ADDITIONAL READING

Bicalho, K.V., Marinho, F.A., Fleureau, J.M., Correia, A.G., and Ferreira, S., 2018. Evaluation of Filter Paper Calibrations for Indirect Determination of Soil Suctions of an Unsaturated Compacted Silty Sand. *Proceedings of the 17th International Conference on Soil Mechanics and Geotechnical Engineering*, Vol. 1, pp. 777–780.

Bicalho, K.V., Correia, A.G., Ferreira, S., Fleureau, J.M., and Marinho, F.A., 2007. Filter Paper Method of Soil Suction Measurement. *Proceedings of the 13th Pan American Conference on Soil Mechanics and Geotechnical Engineering*, pp. 737–741.

Bicalho, K.V., Cupertino, K.F., and Bertolde, A.I., 2013. Evaluation of Suction Calibration Curves for Whatman 42 Filter Paper. *Advances in Unsaturated Soils*, pp. 225–230.

Fondjo, A.A., Theron, E., and Ray, R.P., 2020. Assessment of Various Methods to Measure the Soil Suction. *International Journal of Innovative Technology and Exploring Engineering*, Vol. 9, No. 12, pp. 171–178.

Hamblin, A.P., 1981. Filter Paper Method for Routine Measurement of Field Water Potential. *Journal of Hydrology*, Vol. 53, pp. 355–360.

Harrison, B., and Blight, G., 1998. The Effect of Filter Paper and Psychrometer Calibration Techniques on Soil Suction Measurements. *Proceedings of the 2nd International Conference on Unsaturated Soils*, pp. 362–367.

Houston, S., Houston, W., and Wagner, A., 1994. Laboratory Filter Paper Suction Measurements. *Astm Geotechnical Testing Journal*, Vol. 17, No. 2, pp. 185–194.

Leong, E., He, L., and Rahardjo, H., 2002. Factors Affecting the Filter Paper Method for Total and Matric Suction Measurements. *ASTM Geotechnical Testing Journal*, Vol. 25, No. 3, pp. 322–333.

Leong, E., Kizza, R., and Rahardjo, H., 2016. Measurement of Soil Suction Using Moist Filter Paper. *E3S Web of Science*, Vol. 9, pp. 10012–10017.

Likos, W.J., and Lu, N., 2002. Filter Paper Technique for Measuring Total Soil Suction. *Transportation Research Record* No. 1788, pp. 120–128.

McQueen, I.S., and Miller, R.F., 1968. Calibration and Evaluation of a Wide-Range Gravimetric Method for Measuring Moisture Stress. *Soil Science*, Vol. 106, No. 3, pp. 225–231.

Munoz-Castelblanco, J.A., Pereira, J.M., Delage, P., and Cui, Y.J., 2010. Suction Measurements on a Natural Unsaturated Soil: An Appraisal of the Filter Paper Method. *Proceedings of UNSAT 2010*, pp. 707–712.

Rojas, J.C., Pagano, L., Zingariello, M.C., Mancuso, C., Giordano, G., and Passeggio, G., 2008. A New High Capacity Tensiometer: First Results. *Proceedings of the 1st European Conference on Unsaturated Soils*.

Chapter 22

Thermal Conductivity

INTRODUCTION

The thermal properties of soils can be an important parameter in the design of heat storage systems, ground source heat systems, high level nuclear waste isolation, energy piles, thermal ground improvement techniques and buried pipelines. The Thermal Conductivity of soils depends on a number of factors, including grain-size distribution, soil composition, salt content, Organic Matter, Water Content and Voids Ratio. Several empirical methods are available to estimate Thermal Conductivity of soils from various compositional parameters; however, there is a simple laboratory method that can be used to measure Thermal Conductivity directly.

Unlike some materials, such as water, steel and aluminum, the Thermal Conductivity of soils is not a material property. Since soils consist of mineral matter, water and sometimes air, the Thermal Conductivity depends on composition (grain size and mineralogy), Unit Weight or Porosity/Voids Ratio, Water Content and the degree of Saturation. Many studies have been performed in the laboratory and field to determine the Thermal Conductivity for a wide range of soils.

In the laboratory, Thermal Conductivity is determined by a variation of the line source test method using a thermal needle having a large length-to-diameter ratio to simulate an infinitely long heat source. The test procedure described by ASTM was presented by Chaney et al. (1983) which was a modification of the procedure introduced by Winterkorn (1970). Figure 22.1 shows a typical arrangement for laboratory tests to determine Thermal Conductivity. There are also several commercial probes and read-out devices that can be purchased.

For an infinitely long heat source, if a constant power is applied, the temperature increase, ΔT, at some time, t, after the start of heating is directly related to the Thermal Conductivity. A semi-log plot of T vs. t will give a straight line after some short initial heating period (1–2 min). The linear portion of the plot is used to calculate Thermal Conductivity.

DOI: 10.1201/9781003263289-24

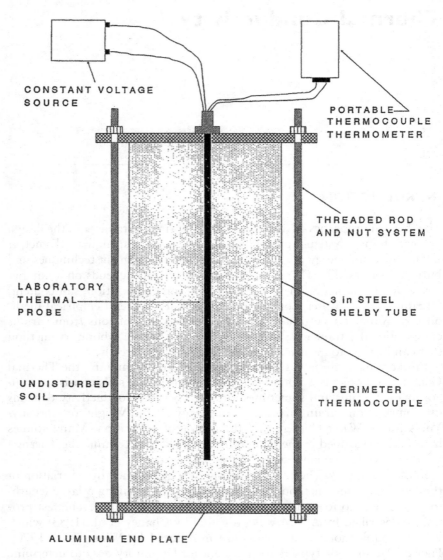

Figure 22.1 Schematic of typical specimen fixture (thin wall Shelby tube) and test arrangement for measuring Thermal Conductivity of undisturbed soils (Lutenegger & Lally 2001).

STANDARD

ASTM D5334 *Standard Test Method for Determination of Thermal Conductivity of Soil and Soft Rock by Thermal Needle Procedure*

EQUIPMENT

Thermal needle (ASTM D5334 describes a technique for constructing
   a suitable thermal needle. Alternatively, thermal needles may be pur-
   chased from several manufacturers.)
Constant electrical current source
Temperature readout or recorder
Digital voltage-ohm-meter capable of reading voltage and current to
   nearest 0.1 volt and ampere
Stopwatch
Thermal grease
Specimen fixture or mold on the order of 180 mm (7 in) in length
Drill and drill bit capable of drilling a straight hole with a diameter
   slightly less than the diameter of the thermal needle and with suffi-
   cient length at least equal to the length of the thermal needle

CALIBRATION OF THERMAL NEEDLE

Each thermal needle needs to be calibrated before its use. Calibration is performed by performing a Thermal Conductivity test using a material with
known thermal properties. A calibration constant can then be determined
for that needle as:

$$C = \lambda_{\text{KNOWN}} / \lambda_{\text{MEASURED}} \qquad (22.1)$$

Glycerine (glycol) is a good reference material to use for calibration and has
a Thermal Conductivity of 0.292W/mK at 25°C. The calibration is performed using the same procedure presented below for testing soils.

PROCEDURE

1. Select a specimen for testing and trim approximately 6.5 mm (1/4
   in) of soil from both ends of the specimen so that the mold sticks up
   above the specimen.
2. Prepare a 50/50 mixture of paraffin and Vaseline and fill the ends of
   the mold with the mixture. This will prevent the specimen from drying out during the test.
3. Coat the thermal needle with a thin coating of thermal grease.
4. Insert the thermal needle into the center of the specimen, making sure
   that the needle is fully embedded.
5. Allow the specimen and fixture to come to equilibrium (approximately 30 min) at the testing temperature. (Note: This may be at room

temperature if the test is performed in a laboratory room with constant temperature for the test period.)

6. Connect the heater wire of the thermal needle to the constant current source.
7. Connect the thermocouple leads to the readout device or data logger.
8. Apply a known constant current to the heater wire so that the temperature change from test temperature is less than about 6–10°C in 1000 sec (ASTM D5334 recommends 1.0 A.). Start the stopwatch.
9. Record the time and temperature readings for a least 20–30 steps throughout the heating period. (Note: Convenient time intervals for temperature readings are 0, 10, 30 sec; and 1, 2, 4, 8, 16 min)

CALCULATIONS

According to ASTM D5334, Thermal Conductivity is calculated from:

$$\lambda = CQ/(4\pi S) = 2.3CQ/4\pi S_{10} \tag{22.2}$$

where:
λ = Thermal Conductivity (W/mK)
C = calibration constant
Q = heat input (W/m) = EI/L
S = slope of linear portion of test curve if ln(t) is used in analysis
S_{10} = slope of linear portion of test curve if $\log_{10}(t)$ is used in analysis
T = time (sec)
I = current flowing through needle (A)
R = total resistance of needle (Ω)
L = length of needle (m)
E = measured voltage (V)

EXAMPLE

A Thermal Conductivity test was conducted on an undisturbed sample of varved clay collected at a depth of 12.2 m (40 ft) at Hadley, Ma. The 76 mm (3 in) Shelby tube was trimmed to a length of 178 mm (7 in). A thermal needle with a diameter of 3 mm (0.12 in) and a length of 150 mm (6 in) was used. The needle had a net resistance of 25.2 ohms, giving a unit resistance of 1.68 ohms/cm. The heat input for the test was 08.62W/m. A calibration constant obtained using glycerin was C = 0.094. Results of the test are shown in Figure 22.2.

The slope of the linear portion of the test gives a value of $S_{10} = \Delta T/\Delta t = 0.0022$°C/sec. The calculated Thermal Conductivity obtained using Equation 22.2 is 0.88 W/m-K.

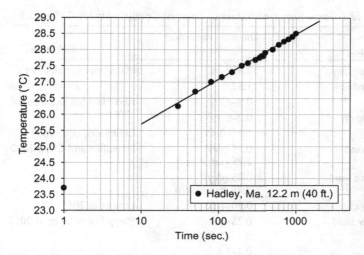

Figure 22.2 Results of Thermal Conductivity test on Varved Clay, Hadley, Ma.

DISCUSSION

Determining the Thermal Conductivity of soils using the needle method is relatively simple and only requires a few specialty pieces of equipment. Several manufacturers have thermal needles and readout devices that can be purchased off the shelf. The test procedure described in this chapter is easy to perform and had been shown to provide reliable results in a wide range of soils, including both fine-grained and coarse-grained materials (e.g., Lutenegger & Lally 2001; Nusier & Abu-Hamdeh 2003; Abuel-Naga et al. 2009; Rozanski & Sobotka 2013; Low et al 2015; Morais et al. 2019). The results of the test depend on good contact between the needle and soil specimen.

The thermal needle should have a minimum length/diameter ratio of 25, but preferably closer to 50. The minimum diameter of the mold should be 51 mm (2 in). Undisturbed specimens may be testing directly in sections of thin-wall tube as illustrated in Figure 22.1. Compacted specimens may be tested in a suitable size mold to fit the thermal needle. For very soft soils, the needle can usually be inserted into the center of the specimen without a pilot hole. In stiff or compacted soils, it will be necessary to use a predrilled hole.

Penner et al. (1975) describe the use of a compaction mold similar to the Shelby tube shown in Figure 22.1. The mold should be sufficiently long so that the thermal needle is fully embedded into the specimen. Once the test is set up, the actual test only takes about 30 min to perform after the initial equilibrium period. After performing the heating test, the cooling response can also be measured to give a secondary determination of Thermal Conductivity. Table 22.1 gives some reported values of Thermal

Table 22.1 Some Reported Values of Thermal Conductivity for Different Soils

Soil type	Thermal Conductivity (W/mK)	Reference
London clay	2.45	Midttomme et al. 1998
Oxford clay	1.57	
Sand	0.9–1.15	Abu-Hamdeh and Reeder
Clay loam	0.2–0.35	2000
Peat	0.05–0.45	O'Donnell et al. 2009
Fine sand	0.15 (dry) 2.75 (saturated)	Hamdhan and Clark 2010
Coarse sand	0.25 (dry) 3.75 (saturated)	
Sandy clay	0.90–1.30	Low et al. 2013
Clayey sand	0.25–3.0	Barry-Macaulay et al. 2013
Sand	0.2–2.5	
Clay	0.2–1.6	
Sand	0.7–2.0	Tong et al. 2016
Loam	0.25–1.5	
Sand	0.2–2.0	Alrtinmi et al. 2016
Silty sand	0.3–0.6 (dry) 2.0–5.5 (saturated)	Roshankhah et al. 2021
Organic silty clay	0.2–0.8	Liu et al. 2020
Clay	1.2–1.5	Lopez-Acosta et al. 2021
Silt	1.–2.0	

Conductivity obtained using the needle method for different soils. To put some of these numbers in perspective, water = 0.6 W/mK; steel = 45 W/mK; aluminum = 237 W/mK.

As noted previously, the Thermal Conductivity of soils depends on a number of factors. Results of detailed laboratory tests have shown that in general, Thermal Conductivity increases with increasing Water Content, increasing Unit Weight and increasing Saturation. The range of values given in Table 22.1 reflect differences in Water Content and Unit Weight. Several empirical models have been suggested that allow an estimation of Thermal Conductivity to be made by taking into account factors such as soil type, Water Content and Unit Weight (e.g., Becker et al. 1992; Gangadhara & Singh 1999; Cote & Konrad 2005; Bertermann et al. (2018).

REFERENCES

Abu-Hamdeh, N.H., and Reeder, R.C., 2000. Soil Thermal Conductivity: Effects of Density, Moisture, Salt Concentration, and Organic Matter. *Soil Science Society of America Journal*, Vol. 64, pp. 1285–1290.

Abuel-Naga, H.M., Bergado, D.T., Bouazza, A., and Pender, M.J., 2009. Thermal Conductivity of Soft Bangkok Clay from Laboratory and Field Measurements. *Engineering Geology*, Vol. 105, pp. 211–219.

Alrtimi, A., Rouainia, M, and Haigh, S., 2016. Thermal Conductivity of a Sandy Soil. *Applied Thermal Engineering*, Vol. 106, pp. 551–560.

Barry-Macaulay, D., Bouazza, A., Singh, R.M., Wang, B., and Ranjith, P.G., 2013. Thermal Conductivity of Soils and Rocks from the Melbourne (Australia) Region. *Engineering Geology*, Vol. 164, pp. 131–138.

Becker, B.R., Misra, A., and Fricke, B.A., 1992. Development of Correlations for Soil Thermal Conductivity. *International Communications in Heat and Mass Transfer*, Vol. 19, No. 1, pp. 59–68.

Bertermann, D., Muller, J., Freitag, S., and Schwarz, H., 2018. Comparison Between Measured and Calculated Thermal Conductivities Within Different Grain Size Classes and Their Related Depths. *Soil Systems*, Vol. 2, No. 50, p. 20.

Chaney, R.C., Ramanjaneya, G., Hencey, G., Kanchanastit, P., and Fang, H.Y., 1983. Suggested Test Method for Determination of Thermal Conductivity of Soil by Thermal-Needle Procedure. *ASTM Geotechnical Testing Journal*, Vol. 6, No. 4, pp. 220–225.

Cote, J., and Konrad, J.M., 2005. A Generalized Thermal Conductivity Model for Soils and Construction Materials. *Canadian Geotechnical Journal*, Vol. 42, pp. 443–458.

Gangadhara Rao, M.V., and Singh, D.N., 1999. A Generalized Relationship to Estimate Thermal Resistivity of Soils. *Canadian Geotechnical Journal*, Vol. 36, pp. 767–773.

Hamdhan, I.N., and Clarke, B.G., 2010. Determination of Thermal Conductivity of Coarse and Fine Sand Soils. *Proceedings of the World Geothermal Conference*.

Liu, X., Cai, G., Congress, S.S., Liu, L., and Liu, S., 2020. Investigation of Thermal Conductivity and Prediction Model of Mucky Silty Clay. *Journal of Materials in Civil Engineering, ASCE*, Vol. 32, No. 8, p. 8.

Lopez-Acosta, N.P., Zaragoza-Cardiel, A.J., and Barba-Galdamez, D.F., 2021. Determination of Thermal Conductivity Properties of Coastal Soils for GSHPs and Energy Geostructure Applications in Mexico. *Energies*, Vol. 14, p. 14.

Low, J.E., Loveridge, F.A., and Powrie, W., 2013. Measuring Soil Thermal Properties for Use in Energy Foundation Design. *Proceedings of the 18th International Conference on Soil Mechanics and Foundation Engineering, Technical Committee 307*, pp. 3375–3378.

Low, J.E., Loveridge, F.A., Powrie, W., and Nicholson, D., 2015. A Comparison of Laboratory and In Situ Methods to Determine Soil Thermal Conductivity for Energy Foundations and Other Ground Heat Exchanger Applications. *Acta Geotechnica*, Vol. 10, pp. 209–218.

Lutenegger A.J., and Lally, M.J., 2001. In Situ Measurement of Thermal Conductivity in a Soft Clay. *Proceedings of the International Conference on In Situ Measurement of Soil Properties and Case Histories, Bali*, pp. 391–396.

Midttomme, K., Roaldset, E., and Aagaard, P., 1998. Thermal Conductivity of Selected Claystones and Mudstones from England. *Clay Minerals*, Vol. 33, pp. 131–145.

Morais, T.S., Sousa, J.D., and Tsuha, C., 2019. Measurement of Thermal Conductivity of Unsaturated Tropical Soils by a Needle Probe Method. *Geotechnical Engineering in the XXI Century*, pp. 2379–2387.

Nusier, O.K., and Abu-Hamdeh, N.H., 2003. Laboratory Techniques to Evaluate Thermal Conductivity for Some Soils. *Hat and Mass Transfer*, Vol. 39, pp. 119–123.

O'Donnell, J.J., Romanovsky, V.E., Harden, J.W., and McGuire, A.D., 2009. The Effect of Moisture Content on the Thermal Conductivity of Moss and Organic Soil Horizons from Black Spruce Ecosystems in Interior Alaska. *Soil Science*, Vol. 174, No. 12, pp. 646–651.

Penner, E., Johnston, G.H., and Goodrich, L.E., 1975. Thermal Conductivity Studies of Some Mackenzie Highway Soils. *Canadian Geotechnical Journal*, Vol. 12, No. 3, pp. 271–288.

Roshankhah, S., Garcia, A.V., and Santamarina, J.C., 2021. Thermal Conductivity of Sand-Silt Mixtures. *Journal of Geotechnical and Geoenvironmental Engineering, ASCE*, Vol. 147, No. 2, p. 9.

Rozanski, A., and Sobotka, M., 2013. On the Interpretation of the Needle Probe Test Results: Thermal Conductivity Measurement of Clayey Soils. *Studia Geotechnica et Mechanica*, Vol. 35, No. 1, pp. 195–207.

Tong, B., Gao, Z., Horton, R., Li, Y., and Wang, L., 2016. An Empirical Model for Estimating Soil Thermal Conductivity from Soil Water Content and Porosity. *Journal of Hydrometeorology*, Vol. 17, pp. 601–613.

Winterkorn, H.F., 1970. Suggested Method of Test for Thermal Resistivity of Soil by the Thermal Probe. *ASTM STP*, Vol. 479, pp. 264–270.

ADDITIONAL READING

Abu-Hamdeh, N.H., and Reeder, R.C., 2000. Soil Thermal Conductivity: Effects of Density, Moisture, Salt Concentration and Organic Matter. *Soil Science Society of America Journal*, Vol. 64, pp. 1285–1290.

Brandon, T.L., and Mitchell, J.K., 1989. Factors Influencing Thermal Resistivity of Sands. *Journal of Geotechnical Engineering, ASCE*, Vol. 115, No. 12, pp. 1683–1698.

Clarke, B.G., Agab, A., and Nicholson, D., 2008. Model Specification to Determine Thermal Conductivity of Soils. *Proceedings of the Institution of Civil Engineers – Geotechnical Engineering*, Vol. 161, pp. 161–168.

Dao, L.Q., Delage, P., Tang, A.N., Pereira, J.M., Li, X.L., and Sillen, X., 2014. Anisotropic Thermal Conductivity of Natural Boom Clay. *Applied Clay Science*, Vol. 101, pp. 282–287.

Hanson, J.L., Edil, T.B., and Yesiller, N., 2000. Thermal Properties of High Water Content Materials. *ASTM STP*, Vol. 1374, pp. 137–151.

Mitchell, J.K., and Kao, T.C., 1978. Measurement of Soil Thermal Resistivity. *Journal of the Geotechnical Engineering Division, ASCE*, Vol. 104, No. 10, pp. 1307–1320.

Moench, A.F., and Evans, D.D., 1970. Thermal Conductivity and Diffusivity of Soil Using a Cylindrical Heat Source. *Soil Science Society of America Proceedings*, Vol. 34, pp. 377–381.

Salomone, L.A., and Kovacs, W.D., 1984. Thermal Resistivity of Soils. *Journal of Geotechnical Engineering, ASCE*, Vol. 110, No. 3, pp. 375–389.

Sepaskhah, A.R., and Boersma, L., 1979. Thermal Conductivity of Soils as a function of Temperature and Water Content. *Soil Science Society of America Journal*, Vol. 43, No. 3, pp. 439–444.

Tang, A.M., Cui, Y.J., and Le, T.T., 2008. A Study on the Thermal Conductivity of Compacted Bentonites. *Applied Clay Science*, Vol. 41, pp. 181–190.

Chapter 23

Electrical Resistivity

INTRODUCTION

Electrical Resistivity of soils can be important in terms of evaluating corrosion potential of soil, estimating Water Content and other soil characteristics and cathodic protection of buried structures. Electrical Resistivity is a measure of the ease at which electricity will flow through a material and is based on Ohm's Law which states that the current between two points of a conducting material is proportional to the voltage

$$I = V/R \quad \text{or} \quad R = V/I \tag{23.1}$$

where:
    I = current (amperes)
    V = voltage (volts)
    R = resistance (ohms)

Resistance is not a material property. Resistivity, ρ, takes into account the area of the conductor and the length over which the current passes:

$$\rho = R(A/L) \tag{23.2}$$

A long length of copper wire has a higher resistance than a short length of the same diameter wire, but the resistivity of the copper is the same. Resistivity is the inverse of conductivity. If we say that a material, such as copper, has a very low resistivity, then it means that it has a high conductivity. The traditional units used for resistivity are Ωm or Ωcm ($1\ \Omega m = 100\ \Omega cm$). The Electrical Resistivity of soils ranges over four orders of magnitude, from about $1\ \Omega m$ to $5{,}000\ \Omega m$.

Resistivity, ρ, is a material property for most materials, such as copper, aluminum, steel, etc. For soils, Resistivity is not considered a property because of the three-phase nature of soil and differences in the percentages of solids, fluid and air. Differences in the solid phase composition, i.e., clay vs. sand, clay mineralogy, Organic Content, etc. can give large differences

DOI: 10.1201/9781003263289-25

in Electrical Resistivity (e.g., Kouchi et al. 2019). For example, Figure 23.1 shows the variation in Electrical Resistivity for two different fine-grained soils with Water Content and Porosity.

The Electrical Resistivity can also change, for example, if the composition of the pore fluid changes (e.g., Hutchinson 1961). This was described in Chapter 11, where the electrical conductivity of soil pore fluid was related to Sensitivity. Therefore, Electrical Resistivity is a soil characteristic.

Tests may be performed in the field and are sometimes used during site investigations to delineate differences in the stratigraphy or soil layering at a site or for mapping contaminant plumes, for example from a landfill. In the laboratory, Electrical Resistivity may be performed on undisturbed, remolded or compacted specimens. In many respects, Electrical Resistivity is analogous to Thermal Conductivity, which measured the ease at which heat flows through soil (Chapter 22). Many of the same factors that influence soil Thermal Conductivity also influence Electrical Resistivity.

In the laboratory, the measurement of Electrical Resistivity is relatively straightforward and only requires minimal equipment. An electrical current is passed through a specimen, and the voltage between two points of known distance apart is measured. A schematic of a typical arrangement is shown in Figure 23.2. Usually, a container, such as shown in Figure 23.3, is needed to hold the specimen. In addition, a power supply and a voltmeter are all that is needed.

There are several different methods that can be used to determine Electrical Resistivity, all of which use the same basic principle. The procedures described in this chapter generally follow the recommended procedures given in ASTM G57 and G187. A number of state highway departments

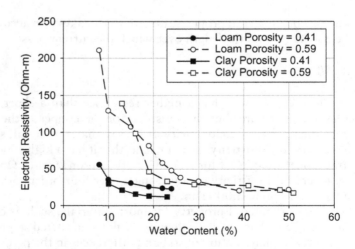

Figure 23.1 Variation in Electrical Resistivity with Water Content and Porosity (data from Seladji et al. 2010).

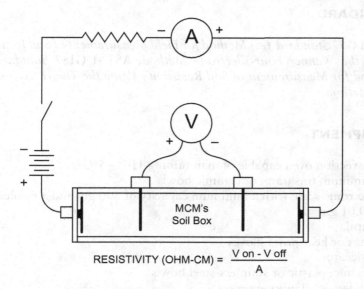

$$\text{RESISTIVITY (OHM-CM)} = \frac{V \text{ on} - V \text{ off}}{A}$$

Figure 23.2 Basic principle used in determining soil Electrical Resistivity.

Figure 23.3 Miller soil box for determining Electrical Resistivity.

have developed test procedures for determining Electrical Resistivity (e.g., Florida, Louisiana) and essentially follow the procedures described in ASTM G57. Additionally, British Standard BS 1377-3 10 describes three methods for determining Electrical Resistivity in the laboratory which are all consistent with ASTM G57.

STANDARD

ASTM G57 *Standard Test Method for Field Measurement of Soil Resistivity Using the Wenner Four-Electrode Method*; ASTM G187 *Standard Test Method for Measurement of Soil Resistivity Using the Two-Electrode Soil Box Method.*

EQUIPMENT

Convection oven capable of maintaining $110° \pm 5°C$
Aluminum tins (tare) or ceramic bowls
Electronic scale with a minimum capacity of 500 gm and a resolution of 0.01 gm
Spatula
Tongs or heat-proof gloves
Desiccator
Ceramic, plastic or stainless-steel bowls
Stainless-steel mixing spoon
Soil box (non-conducting material)
Power supply
Voltmeter

(Note: A common style of soil box for use in determining Electrical Resistivity of soils is a "Miller" soil box, available commercially. However, soil boxes may be fabricated very simply from any non-conducting material, e.g., Plexiglass, Perspex, acrylic or PVC. Miller soil boxes are available as either a small box or large box and have the following characteristics:

Small box
Cross-sectional area = 3 cm × 2.4 cm = 7.2 cm^2
Length between electrodes = 7.2 cm
A/L = 1 cm
Large box
Cross-sectional area = 4 cm × 3.2 cm = 12.8 cm^2
Length between electrodes = 12.8 cm
A/L = 1 cm

Because the ratio of area to length (A/L) is equal to 1 for both boxes, the resistivity, ρ, is numerically equal to the measured resistance).

CALIBRATION

In some cases, it may be necessary to calibrate the soil box or cell using materials with known Electrical Resistivity. This is done using the procedure described below for both the four-electrode and two-electrode methods. Typically, standard solutions of distilled water and sodium chloride (NaCl) having electrical resistivities of 1000, 5000 and 10,0000 Ωcm are used for calibration.

PROCEDURE – FOUR ELECTRODE METHOD

1. Select a representative soil sample for testing.
2. Prepare the specimen as needed for the project. Add water to achieve a target Water Content if needed or test the specimen at natural Water Content. Compacted specimens (e.g., Proctor compaction) may also be tested using a special cell to hold the specimen, or specimens may be trimmed from the compacted sample. (Note: When possible, use groundwater from the sample excavation to add water to the specimen. Otherwise use distilled water. If the soil resistivity is expected to be below 10,000 Ωcm, ASTM G57 allows the use of tap water.)
3. Select a clean dry soil box or cell of known inside dimensions and determine the mass of the dry box or cell. Depending on the design of the soil box, it may be necessary to remove the electrode prior to placing the soil in the box. Some boxes are equipped with end plates rather than pin electrodes.
4. For undisturbed samples, carefully trim a specimen to fit tightly in the soil box or cell for remolded or compacted soils, making sure to fill the box leaving no air gaps. (Note: Some soils may be poured or tamped directly into the box; e.g., testing soil slurries.) Obtain the total mass of the soil box and soil.
5. If electrodes have been removed, replace the electrodes, making sure of good contact between the electrodes and soil.
6. Attached connecting wires from the current source to the outer electrodes and connecting wires from the voltmeter to the inner electrodes. (Note: An AC current source, usually around 97 Hz, is preferred since the use of DC current may cause polarization of the metal electrodes over time; however, for short duration tests, a DC source may also be used. A suitable power supply should be capable of providing 12 to 20 v at about 100 ma.)
7. Apply the current and read the voltage on the inner electrodes. (Note: several four-pine commercial resistivity meters are available: e.g., MC Miller Model 400A, Nilsson Model 400, Tinker & Rasor Model SR-2, etc.)
8. After completing the test, obtain a representative sample of the soil and obtain the Water Content using the convection oven method described in Chapter 1.
9. Thoroughly clean the box with distilled water and dry in preparation for the next test.

CALCULATIONS

Using the measured voltage and resistance and the appropriate electrode spacing, use Eq. 23.2 to calculate the resistivity

$$\rho = R(A/L)$$

(Note: If using a Miller soil box, the resistivity is numerically equal to the resistance value.)

If the current electrodes are not spaced at the same interval as the potential-measuring electrodes, calculate the resistivity as:

$$\rho(\Omega cm) = \pi bR / \left[1 - (b/(b+a))\right] \qquad (23.3)$$

where:
   b = outer electrode spacing (cm)
   a = inner electrode spacing (cm)
   R = resistance (Ω)

Report the soil resistivity to the nearest whole number in either Ωm or Ωcm. Also calculate the dry Unit Weight of the specimen using the measured inside dimensions of the soil box and the mass of the empty and full soil box. Report the Water Content determined from the oven-dried specimen.

PROCEDURE – TWO ELECTRODE METHOD:

The test procedure using the two-electrode method is the same as described above for the four-electrode method, except that the current and potential are measured on the same (outer) electrodes.

CALCULATIONS

If using the two-electrode method, calculate the resistivity as:

$$\rho = R(A/L)$$

For example, if using a cubical soil box with inside dimensions of 4 cm × 4 cm × 4 cm:

$$\rho = (4 cm \times 4 cm)R / 4 cm = 4R(\Omega cm)$$

DISCUSSION

Normally, a rectangular ("Miller" style) soil box as described above, fabricated from a non-conducting material, is used in determining Electrical Resistivity. However, other styles of cells have been used to determine Electrical Resistivity, depending on the type of specimen being tested. Figures 23.4–23.7 show some variations in the geometry of cells reported for both four-electrode and two-electrode methods.

An alternative to using a soil box or cell with internal electrodes is to use a similar nonconducting fixture or container to hold the soil specimen. A small plastic strip holding four electrodes spaced equidistantly to form a Wenner array is then used. The electrodes are slowly pushed into the specimen and the test is performed as usual. This approach can also be used in the field on soil specimens retrieved from sampling or may be used on the open face of excavations or road cuts (e.g., Kaufhold et al. 2014).

Other types of cells, including modified oedometer cells, have also been used successfully to determine Electrical Resistivity (e.g., McCarter & Desmazes 1997; Fukue et al. 1999.; Sreedeep et al. 2004; Seladji et al. 2010; Bhatt & Jain 2014; Vincent et al. 2017; Kibria & Hossain 2015; Choo et al. 2022). All of these different cells appear to be equally suited to performing

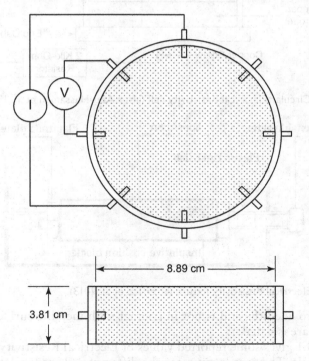

Figure 23.4 Circular Electrical Resistivity cell (after Kalinski & Kelly 1994).

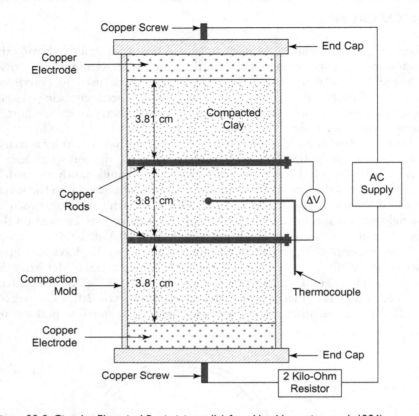

Figure 23.5 Circular Electrical Resistivity cell (after Abu-Hassanien et al. 1996).

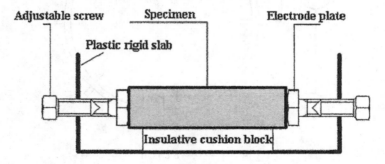

Figure 23.6 Electrical Resistivity fixture (after Bai et al. 2013).

the test, provided the soil is not in contact with metallic surfaces and the electrodes are isolated.

Table 23.1 gives some reported values of Electrical Resistivity for a wide range of soils. The ranges given for the different soil types relate to differences in density, Water Content and Saturation. For reference, copper has a

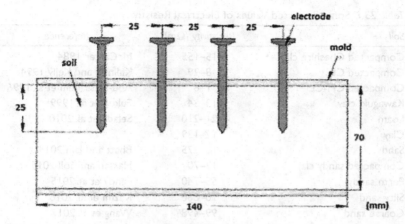

Figure 23.7 Electrical Resistivity mold (after Gingine et al. 2016).

resistivity of about 1.7×10^{-8} Ωm, platinum of about 10.6×10^{-8} Ωm, wood about 10^3 Ωm and seawater 2.1×10^{-1} Ωm.

In general, Electrical Resistivity for a given soil decreases with increasing degree of Saturation, decreases with increasing Water Content and decreases with increasing dry Unit Weight. Fine-grained soils tend to have lower Electrical Resistivity than coarse-grained soils; however, the resistivity of the pore fluid may have considerable influence on resistivity.

Soil Electrical Resistivity measurements are often used independently or in conjunction with pH to estimate the corrosion potential of ferrous metals in soil. A number of field and laboratory studies beginning in the early 1950s have shown a direct relationship between soil Electrical Resistivity and section loss of uncoated steel piles, anchors, plates and pipes (e.g., Schwerdtfeger 1965; Chaker 1981; Sing et al. 2013; Daku et al. 2019). Several corrosion rating systems have been suggested based solely on soil Electrical Resistivity. Table 23.2 gives a commonly used corrosion rating from Roberge (2007). Table 23.3 gives suggested ranges of the degree of corrosion potential related to Electrical Resistivity from the National Association of Corrosion Engineers (NACE) and ASTM G187 for the two-electrode method which are slightly different than the values given in Table 23.2.

A similar corrosivity scale was suggested by Bradford (2000), which also includes an estimate of the years to penetrate sheet steel, as shown in Table 23.4.

The main problem with using these corrosion categories is that they only consider one parameter, i.e., resistivity. The European Standard EN 12501-2:2003 uses a corrosion severity rating that includes both soil pH and resistivity, as given in Table 23.5.

As noted in the introduction, many of the same factors that influence soil Thermal Conductivity also influence the Electrical Resistivity of

Table 23.1 Some Reported Values of Electrical Resistivity

Soil	Resistivity (Ωm)	Reference
Compacted Cheshire clay	15–155	McCarter 1984
Compacted CL	8–29	Kalinski and Kelly 1994
Compacted CH-CL	1–8	Abu-Hassanein et al. 1996
Kawaguki clay	13–34	Fukue et al. 1999
Loam	21–210	Seladji et al. 2010
Clay	12–139	
Sand	10–275	Bhatt and Jain 2014
Compacted sandy clay	12–70	Hassan and Toll 2015
Perth sandy	55–740	Pandy et al. 2015
Silty sand	10–860	Kazmi et al. 2016
Coarse sand	99–498	Wang et al. 2017
Fine sand	46–350	
Medium sand	46–321	
Clay	32–89	
Silt	24–297	
Silty clay	32–248	
ML clayey silt	520–1400	Zohra-Hadjadj et al. 2019
Compacted kaolinite	10–300	Kibria and Hossain 2017
Compacted bentonite	2–22	
Compacted illite	3–44	
SM-SC	7–65	Adarmanahadi et al. 2021

Table 23.2 Severity of Corrosivity of Soils (Roberge 2007)

Resistivity (Ωcm)	Corrosion rating
> 20,000	Essentially non-corrosive
10,000–20,000	Mildly corrosive
5000–10,000	Moderately corrosive
3000–5000	Corrosive
1000–3000	Highly corrosive
< 1000	Extremely corrosive

soils. Several studies have shown direct correlations between Thermal Conductivity and Electrical Resistivity (e.g., Lowell 1984; Singh et al. 2001; Wang et al. 2017). Figure 23.8 shows results of some laboratory measurements on a range of soils.

Electrical Resistivity measurements are also used for site characterization. The Wenner four-electrode method has been adapted to the cone penetration test (CPT) and has been used extensively during site investigations, especially for detecting the extent of contaminated groundwater and pore water salinity (e.g., Campanella & Weemes 1990; Auxt & Wright 1995;

Table 23.3 Severity of Corrosivity of Soils; NACE and ASTM G187

Resistivity (Ωcm)	Corrosion potential	
	NACE	ASTM G187
> 10,000	Negligible	Very mildly corrosive
5001–10,000	Mildly corrosive	Mildly corrosive
2001–5000	Mildly corrosive	Moderately corrosive
1001–2000	Moderately corrosive	Severely corrosive
501–1000	Corrosive	Extremely corrosive
0–500	Very corrosive	Extremely corrosive

Table 23.4 Corrosion Assessment Table from Bradford (2000)

Resistivity (Ωcm)	Steel corrodibility	Years to penetrate sheet steel
> 10,000	Slight to none	> 20
5000–10,000	Slight	> 20
2000–5000	Moderate to slight	15–20
1000–2000	Severe to moderate	10–15
500–1000	Severe	5–10
< 500	Very severe	1–5

Table 23.5 EN 12501-2:2003 Corrosivity Rating

Soil pH	Resistivity (Ωcm)	Corrosion rating
< 3.5	All	High
3.5–4.5	< 4500	High
	> 4500	Medium high
4.5–5.5	< 4500	High
	4000–5000	Medium high
	> 5000	Medium
5.5–6.0	< 1000	High
	1000–5000	Medium high
	5000–10,000	Medium
	> 10,000	Medium low
6.0–9.5	< 1000	High
	1000–3000	Medium high
	3000–10,000	Medium
	10,000–20,000	Medium low
	> 20,000	Low

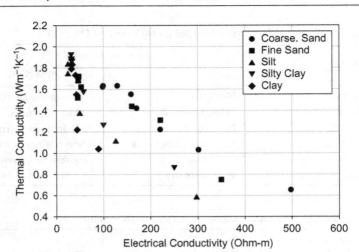

Figure 23.8 Relationship between Electrical Resistivity and Thermal Conductivity for a range of different soils (data from Wang et al. 2017).

Long et al. 2012; Cai et al. 2016). Figure 23.9 shows some typical CPT probe modules used to measure Electrical Resistivity.

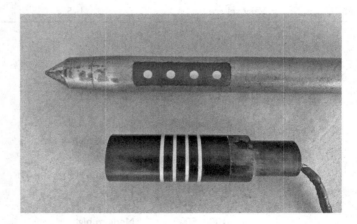

Figure 23.9 CPT Electrical Resistivity modules.

REFERENCES

Abu-Hassanein, Z.S., Benson, C.H., and Blotz, L.R., 1996. Electrical Resistivity of Compacted Clays. *Journal of Geotechnical Engineering, ASCE*, Vol. 122, No. 5, pp. 397–406.

Adarmanahadi, H.R., Rasti, A., and Razavi, M., 2021. The Effects of Kiln Dust on the Soil Electrical Resistivity. *American Journal of Engineering and Applied Sciences*. Vol. 14, No. 1, pp. 51–63.

Auxt, J.A., and Wright, D., 1995. Environmental Site Characterization in the U.S. Using the Cone Penetrometer. *Proceedings of the International Symposium on Cone Penetration Testing*, Vol. 2, pp. 387–392.

Bai, W., Kong, L., and Guo, A., 2013. Effects of Physical Properties on Electrical Conductivity of Compacted Lateritic Soil. *Journal of Rock Mechanics and Geotechnical Engineering*, Vol. 5, pp. 406–411.

Bhatt, S., and Jain, P.K., 2014. Correlation Between Electrical Conductivity and Water Content of Sand – A Statistical Approach. American International Journal of Research in Science, Technology. *Engineering and Mathematics*, Vol. 6, No. 2, pp. 115–121.

Bradford, S.A., 2000. *Practical Handbook of Corrosion Control in Soils: Pipelines, Tanks, Casings, Cables*. Casti Publishing: Edmonton, Canada.

Cai, G., Chu, Y., Liu, S., and Puppala, A.J., 2016. Evaluation of Subsurface Spatial Variability in Site Characterization Based on RCPTU Data. *Bulletin of Engineering Geology and the Environment*, Vol. 75, pp. 401–412.

Campanella, R.G., and Weemes, I., 1990. Development and Use of an Electrical Resistivity Cone for Groundwater Contamination Studies. *Canadian Geotechnical Journal*, Vol. 27, pp. 557–567.

Chaker, V., 1981. Simplified Methods for the Soil Electrical Resistivity Measurement. *ASTM STP*, Vol. 741, pp. 61–91.

Choo, H., Park, J., Do, T.T., and Lee, C., 2022. Estimating Electrical Conductivity of Clayey Soils with Varying Mineralogy Using the Index Properties of Soils. *Applied Clay Science*, Vol. 217, p. 9.

Daku, S.S., Diyelmak, V.B., Oyitolaiye, O.A., and Abalaka, I.E., 2019. Evaluation of Soil Corrosivity Using Electrical Resistivity Method: A Case Study of Part of the University of JOS Permanent Site. *Scientific Research Journal (SCIRJ)*, Vol. 7, No. 3, pp. 66–74.

Fukue, M., Minato, T., Horibe, H., and Taya, N., 1999. The Micro-Structure of Clay Given by Resistivity Measurements. *Engineering Geology*, Vol. 54, pp. 43–53.

Gingine, V., Dias, A.S., and Cardoso, R., 2016. Compaction Control of Clayey Soils Using Electrical Resistivity. *Procedia Engineering*, Vol. 143, pp. 803–810.

Hassan, A.A., and Toll, D.G., 2015. Water Content Characteristics of Mechanically Compacted Clay Soil Determined Using Electrical Resistivity Method. *Proceedings of the 16th European Conference on Soil Mechanics and Geotechnical Engineering*, pp. 3395–3400.

Hutchinson, J.N., 1961. A Landslide on a Thin Layer of Quick Clay at Furre, Central Norway. *Geotechnique*, Vol. 11, pp. 69–94.

Kalinski, R.J., and Kelly, W.E., 1994. Electrical-Resistivity Measurements for Evaluating Compacted-Soil Liners. *Journal of Geotechnical Engineering, ASCE*, Vol. 120, No. 2, pp. 451–457.

Kaufhold, S., Grissemenn, C., Dohrmann, R., Klinkenberg, M., and Decher, A., 2014. Comparison of Three Small-Scale Devices for the Investigation of the Electrical Conductivity/Resistivity of Swelling and Other Clays. *Clays and Clay Minerals*, Vol. 62, No. 1, pp. 1–12.

Kazmi, D., Qasim, S., Siddiqui, F.I., and Azhar, S.B., 2016. Exploring the Relationship Between Moisture Content and Electrical Resistivity for Sandy and Silty Soil. *International Journal of Engineering science Invention*, Vol. 5, No. 7, pp. 42–47.

Kibria, G., and Hossain, S., 2017. Electrical Resistivity of Compacted Clay Minerals. *Environmental Geotechnics*, Vol.6, No. 1, pp. 19–25.

Kouchaki, B.M., Bernhardt-Barry, M.L., Wood, C.M., and Moody, T., 2019. A Laboratory Investigation of Factors Influencing the Electrical Resistivity of Different Soil Types. *ASTM Geotechnical Testing Journal*, Vol 42, No. 4, pp. 829–853.

Long, M., Donohue, S., L'Heureux, J.S., Solberg, I.L., Renning, J.S., Limcher, R., O'Conner, P., Sauvi, G., Remoen, M., and Lecomte, I., 2012. Relationship Between Electrical Resistivity and Basic Geotechnical Parameters for Marine Clays. *Canadian Geotechnical Journal*, Vol. 49, pp. 1158–1168.

Lowell, M.A., 1984. Thermal Conductivity and Permeability Assessment by Electrical Resistivity Measurements in Marine Sediments. *Marine Geotechnology*, Vol. 6, pp. 205–239.

McCarter, W.J., 1984. The Electrical Resistivity Characteristics of Compacted Clays. *Geotechnique*, Vol. 34, No.2, pp. 263–267.

McCarter, W.J., and Desmazes, P., 1997. Soil Characterization Using Electrical Measurements. *Geotechnique*, Vol. 47, No. 1, pp. 179–183.

Pandy, L.M., Shukla, S.K., and Habibi, D., 2015. Electrical Resistivity of Sandy Soil. *Geotechnique Letters*, Vol. 5, pp. 175–185.

Schwerdtfeger, W.J., 1965. Soil Resistivity as Related to Underground Corrosion and Cathodic Protection. *Journal of Research of the National Bureau of Standards*, Vol. 69C, No. 1, pp. 71–77.

Seladji, S., Cosenza, P., Tabbagh, A., Ranger, J., and Richard, G., 2010. The Effect of Compaction on Soil Electrical Resistivity: a Laboratory Investigation. *European Journal of Soil Science*, Vol. 61, No. 6, pp. 1043–1055.

Sing, L.K., Yahaya, N., Othmman, S.R., Fariza, S.N., and Noor, N.M., 2013. Resistivity and Corrosion Growth in Tropical Region. *Journal of Corrosion Science and Engineering*, Vol. 16, p. 11.

Singh, D.N., Kuriyan, S.J., and Manthena, K.C., 2001. A Generalized Relationship Between Soil Electrical and Thermal Resistivities. *Experimental Thermal and Fluid Science*, Vol. 25, pp. 175–181.

Sreedeep, S., Reshma, A.C., and Singh, D.N., 2004. Measuring Soil Electrical Resistivity Using a Resistivity Box and Resistivity Probe. *ASTM Geotechnical Testing Journal*, Vol. 27, No. 4, pp. 411–415.

Vincent, N.M., Shivasankar, R., and Lokesh, K.N., 2017. Some Studies on Laboratory and Field Electrical Resistivities of Soils. *Proceedings of the Indian Geotechnical Conference*.

Wang, J., Zhang, X., and Du, L., 2017. A Laboratory Study of the Correlation Between Thermal Conductivity and Electrical Resistivity of Soil. *Journal of Applied Geophysics*, Vol. 145, pp. 12–16.

ADDITIONAL READING

Arriba-Rodriguez, L., Villanueva-Balsera, J., Ortega-Fernandez, F., and Rodriguez-Perez, F., 2018. Methods to Evaluate Corrosion in Buried Steel Structures: A Review. *Metals*, Vol. 8, No. 334, p. 21.

Daniel, C.R., Campanella, R.G., Howie, J.A., and Giacheti, H.L., 2003. Specific Depth Cone Resistivity Measurements to Determine Soils Engineering

Properties. *Journal of Environmental Engineering and Geophysics*, Vol. 8, No. 1, pp. 15–22.

Gupta, S.C., and Hanks, R.J., 1972. Influence of Water Content on Electrical Conductivity of the Soil. *Soil Science Society of America Proceedings*, Vol. 36, No. 6, pp. 855–857.

Hazrek, Z.A., Aziman, M., Azhar, A.T., Cloitral, W., Fauziah, A., and Rosli, S., 2015. The Behavior of Laboratory Soil Resistivity Value Under Basic Soils Properties Influence. *IOP Conference Series: Earth and Environmental Science*, Vol. 23, Series 012002, p. 9.

Owusu-Nimo, F., Peprah-Manu, D., Kumah, D., and Ampadu, S. I., 2019. Compaction Verification of Lateritic Soils Using Electrical Resistivity and Thermal Conductivity. *Proceedings of the 17th African Regional Conference on Soil Mechanics and Geotechnical Engineering*, pp. 339–344.

Rhoades, J.D., Raats, P.A., and Prather, R.J., 1976. Effect of Liquid-Phase Electrical Conductivity, Water Content and Surface Conductivity on Bulk Soil Electrical Conductivity. *Soil Science Society of America Journal*, Vol. 40, pp. 651–655.

Rhoades, J.D., and Van Shilfgaarde, J., 1976. An Electrical Conductivity Probe for Determining Soil Salinity. *Soil Science Society of America Journal*, Vol. 40, pp. 647–651.

Seladji, S., Cozenza, P., Tabbagh, A., Ranger, J., and Richard, G., 2010. The Effect of Compaction on Soil Electrical Resistivity: A Laboratory Investigation. European *Journal of Soil Science*, p. 14.

Vilda, W.S. III, 2009. *Corrosion in the Soil Environment: Soil Resistivity and pH Measurements*. NCHRP Final Report, Transportation Research Board, National Research Council.

Yan, M., Miao, L., and Cui, Y., 2012. Electrical Resistivity Features of Compacted Expansive Soils. *Marine Georesources and Geotechnology*, Vol. 30, pp. 167–179.

Zohra-Hadjadj, F., Laredj, N., Maliki, M., Missoum, H., and Nemdani, K., 2019. Laboratory Evaluation of Soil Geotechnical Properties via Electrical Conductivity. *Revista Facultad de Ingenieria, Universidad de Antioquia*, Vol. 99, pp. 101–112.

Chapter 24

Index Strength Tests and Sensitivity

INTRODUCTION

A detailed discussion of laboratory test procedures to determine the undrained shear strength of clays and other fine-grained soils is beyond the scope of this manual; however, geotechnical engineers often use simple, inexpensive tests to make initial estimates of strength. These types of tests are often referred to as *strength index tests* because they can be quite crude and are not performed under carefully controlled laboratory conditions. Nonetheless, they actually can give a very good indication of the relative stiffness of different soils, and in some cases the results are quite useful. While these tests are primarily intended for use with undisturbed samples of predominantly fine-grained soils such as silts and clays, they may also be used on compacted fine-grained soils.

Four different laboratory Index Strength Tests are described in this chapter: 1) pocket penetrometer; 2) pocket vane (Torvane); 3) laboratory miniature vane; and 4) Fall Cone. Even though the tests are normally intended for use in the laboratory, the pocket penetrometer and pocket vane may also be performed in the field during a site investigation by testing the ends of trimmed thin-walled samples.

POCKET PENETROMETER

The pocket penetrometer was introduced in the mid-1950s as a simple portable method for estimating the undrained shear strength of clays. An early advertisement for the pocket penetrometer is shown in Figure 24.1. It's possible that the pocket penetrometer was an update of an earlier similar device used in and known as the Danish Railways spring-scale cone (Godskesen 1936) shown in Figure 24.2. The pocket penetrometer consists of a simple plunger device that uses an internal compression spring to determine the relative strength of a soil by forcing the plunger into the soil to a fixed depth and measuring how much the spring compresses.

DOI: 10.1201/9781003263289-26

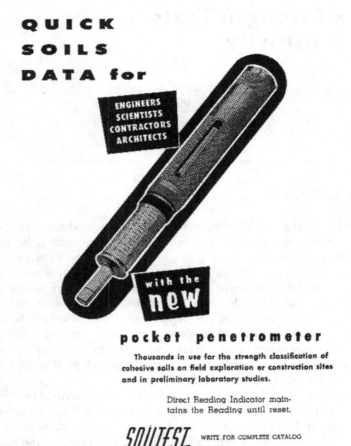

Figure 24.1 Early (circa 1957) advertisement for pocket penetrometer.

The force on the plunger is determined by the compression of the spring. The further the spring compresses, the larger the force, and the higher the strength of the soil. The penetrometer has a plunger diameter of 6.35 mm (0.25 in) and is calibrated from the factory with a scale on the side so that it reads directly in unconfined compressive strength, q_u; in units of kg/cm^2 or tons/ft^2 (tsf). The device is based on the principle of an undrained bearing capacity failure of a circular footing on the surface of clay (Figure 24.3).

Standard

There is currently no ASTM standard test method for this test.

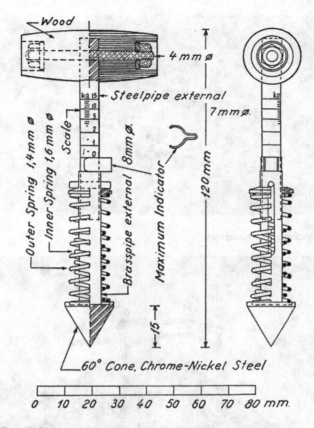

Figure 24.2 Danish Railways spring-scale cone (Godskesen 1936).

Equipment

Convection oven capable of maintaining $110° \pm 5°C$
Aluminum tins (tare) or ceramic bowls
Electronic scale with a minimum capacity of 500 gm and a resolution of
0.01 gm
Spatula
Tongs or heat-proof gloves
Desiccator
Pocket Penetrometer with enlarged foot attachment

Procedure

1. Select a specimen for testing. Make sure that the end of the specimen
is flat.

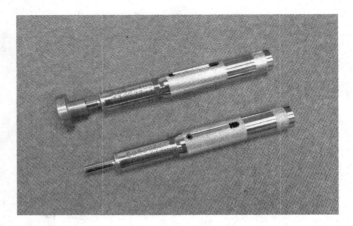

Figure 24.3 Pocket Penetrometers.

Figure 24.4 Conducting pocket penetrometer test on the end of a thin-walled tube sample.

2. Slide the ring down against the handle end of the penetrometer and make sure that it reads ZERO.

3. Grip the body of the penetrometer and slowly push the tip of the Penetrometer into the soil to the groove on the plunger located 3.35 mm (¼ in) from the tip. The Penetrometer should be held perpendicular to the surface being tested and should be pushed straight in at a slow steady rate. Figure 24.4 shows a pocket penetrometer test being conducted on the end of a thin-walled sample.

4. Read and record the pocket penetrometer unconfined compressive strength in kg/cm^2 (tons/ft^2) directly from the scale as indicated by the ring. The reading is taken on the lower side of the ring, that is, the side closest to the handle. Record the size of foot used.

5. Repeat the test two or three times at adjacent locations, leaving about three diameters between test locations, and record all of the readings. (Note: On a 76 mm [3 in] thin-walled tube, usually three or four tests may be performed unless the soil is very soft.)
6. After completing the test, obtain a small soil sample from the test region and determine the Water Content using the convection oven method described in Chapter 1.

A large foot attachment is available for very soft soils if the standard end of the penetrometer does not give a reliable reading, as shown in Figure 24.5. If very soft soils are being tested using the large foot, which has a diameter of 25.4 mm (1 in), the unconfined strength reading from the penetrometer should be multiplied by 16 to obtain the correct value.

One problem with the Pocket Penetrometer is that in very hard soils, the device will reach a limiting value without penetrating the required 0.5 in. In this case, the results are reported as > 4.5 kg/cm^2 (tsf). The author has fabricated an attachment with a smaller size foot for use in very stiff clays, such as glacial tills, shown in Figure 24.6. The small foot has an area half that of the standard penetrometer end. This means that the number obtained from the pocket penetrometer scale should be multiplied by two when using the small foot.

Calculations

There are no calculations for this test. The unconfined compressive strength is read directly from the scale on the body of the penetrometer.

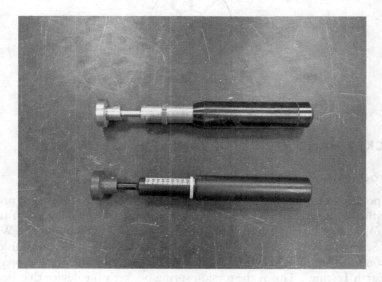

Figure 24.5 Large foot attachment for pocket penetrometer use in very soft soils.

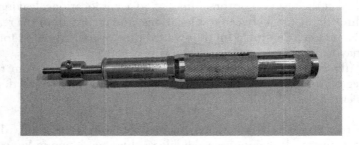

Figure 24.6 Small foot attachment for pocket penetrometer use in very soft soils.

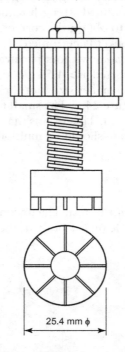

25.4 mm ϕ

Figure 24.7 Schematic of pocket vane (Torvane).

POCKET HAND VANE (TORVANE)

Introduction

The pocket vane test or torvane is another device that may be used to provide an estimate of strength. The torvane was introduced by Sibley and Yamane (1966), who credit the design to Stanley Wilson, then of Shannon and Wilson A U.S. Patent (No. 3,364.734) was issued on Jan. 23, 1968 to Stanley D. Wilson for a small handheld shear device "Torsional Vane Shear Apparatus for Earth Testing". This is the torsion spring pocket vane device that became known commericllay as the Torvane. The device, shown in Figure 24.7, has

Figure 24.8 Torvane kit with different size vanes for different consistency soils.

an internal torsion spring that applies load to the soil through a series of blades on the end of the torvane that are pushed into the soil. The operator rotates the handle to apply torque to the vanes to shear the soil. The head of the torvane is marked off directly to give undrained shear strength in units of kg/cm^2 or $tons/ft^{2.}$ and has a small pointer for directly reading the strength value. For very soft or very hard soils, special vanes may be attached to the bottom to more accurately measure the strength. Figure 24.8 shows a photo of a torvane kit with all three size vanes. Torvanes are available from a number of different equipment supply companies and can be fabricated from mild stainless steel, aluminum or hard plastic.

To perform the test, the dial pointer on the handle is set to zero. After making sure that the blades are clean, the blades are pushed into the end of the soil to be tested until they bottom out. The handle is then slowly rotated clockwise until the soil shears. The shear strength is read directly from the scale on the end of the handle. This procedure is repeated two or three times with all of the readings recorded. The torvane is typically performed on the end of a thin-walled tube sample, as shown in Figure 24.9.

Standard

ASTM D8121 *Standard Test Method for Approximating the Shear Strength of Cohesive Soils by the Handheld Vane Shear Device*

Equipment

Convection oven capable of maintaining $110° \pm 5°C$
Aluminum tins (tare) or ceramic bowls

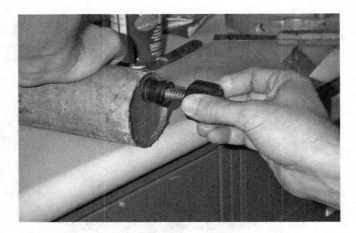

Figure 24.9 Conducting a Torvane test on the end of a thin-walled tube sample.

Electronic scale with a minimum capacity of 500 gm and a resolution of 0.01 gm

Spatula

Tongs or heat-proof gloves

Desiccator

Torvane

Extra vane attachments (for very hard or very soft soils)

Procedure

1. Set the dial pointer on the handle to ZERO.
2. Make sure that the blades are clean and push the torvane into the end of the soil to be tested until it bottoms out.
3. Apply a small normal force to the torvane and slowly rotate the handle clockwise in a smooth motion until the soil shears.
4. Read and record the torvane undrained shear strength, s_u, in kg/cm^2 or Tons/ft^2 directly from the dial pointer. Record the size of vane used.
5. If possible, repeat the readings two or three times, leaving one vane diameter between the test locations and record all of the readings. (Note: On a 76 mm [3 in] thin-walled tube, usually three tests may be performed unless the soil is very soft.)
6. If the remolded strength is desired, thoroughly remold the soil in a plastic bag, kneading by hand. Fill a small plastic cup with the soil, making sure to remove any trapped air and repeat the test, Figure 24.10.
7. After completing the test, obtain a small soil sample from the test region and determine the Water Content using the convection oven method described in Chapter 1.

Figure 24.10 Performing Torvane test on a remolded specimen.

The Torvane kit includes vane ends that can be interchanged for more accurate use in either soft clays or stiff clays, Figure 24.8. If using the larger vane for soft clays, the reading on the torvane must be multiplied by 0.2; if using the small vane for stiff soils, the reading must be multiplied by 2.5. The large vane is especially useful for testing remolded soils so that an indication of Sensitivity can be obtained.

Calculations

There are no calculations to be made; read the strength directly from the scale on the Torvane and adjust as appropriate for the size of the vane used. Remember that the Torvane provides a measure of the undrained shear strength, s_u, while the Pocket Penetrometer provides a measure of unconfined compressive strength, q_u. The value from the Pocket Penetrometer must be divided by two to give comparable undrained shear strength.

LABORATORY VANE

Introduction

A somewhat more sophisticated index test for obtaining an estimate of undrained shear strength is the laboratory vane test. Laboratory vane tests are very similar to field vane tests, except that the vane and equipment are on a smaller scale (miniature) and the test is intended for use on small undisturbed samples. Early hand operated laboratory vanes were used by Evans and Sherratt (1948), Skempton and Bishop (1950), Jakobson (1954) and Gray (1957). A motorized lab vane was used in the early 1960s by Karrlson (1961). An early version of the equipment is shown in Figure 24.11. The equipment has changed little since the 1950s. Both hand-operated and motorized laboratory vane equipment are available.

The test uses torsion springs attached to the head of the vane to measure soil resistance on a four-bladed vane inserted into the soil. The vane is rotated until failure and the measured torque is used to calculate the undrained shear strength, s_u, based on the traditional theory of shearing resistance as used with the field vane test. Different size vanes, typically 25.4 mm × 12.7 mm (1 in × ½ in) or 12.7 mm × 12.7 mm (½ in × ½ in), as shown in Figure 24.12, and different torque springs are available to test soils with

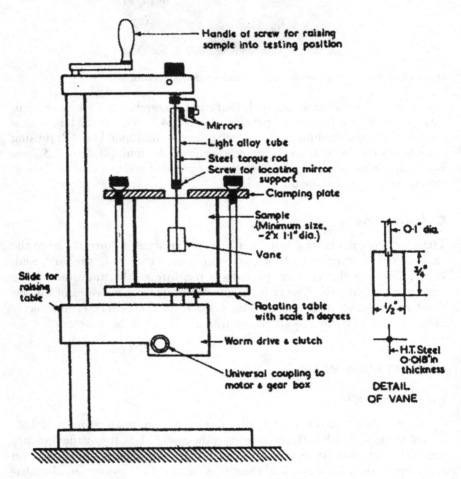

Figure 24.11 Early laboratory miniature vane equipment (from Skempton and Bishop 1950).

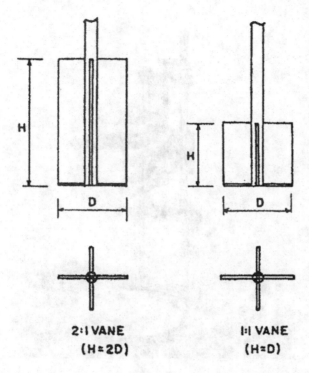

2:1 VANE
(H≈2D)

1:1 VANE
(H=D)

Figure 24.12 Typical sizes of vanes used with laboratory miniature vane test.

different consistencies. Tests may be conducted on both undisturbed and completely remolded samples in order to obtain an estimate of Sensitivity. Figure 24.13 shows a photo of a modern laboratory vane.

Standard

ASTM D4648 *Standard Test Method for Laboratory Miniature Vane Shear for Saturated Fine Grained Clayey Soil*

Equipment

Convection oven capable of maintaining $110° \pm 5°C$
Aluminum tins (tare) or ceramic bowls
Electronic scale with a minimum capacity of 500 gm and a resolution of
 0.01 gm
Spatula
Tongs or heat-proof gloves
Desiccator
Laboratory vane with a selection of torsion springs

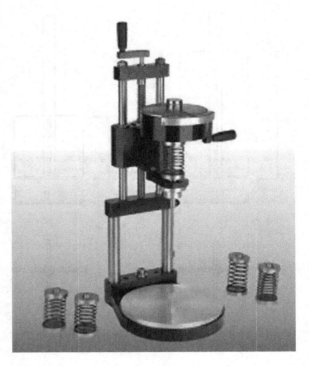

Figure 24.13 Modern miniature laboratory vane test.

Procedure

1. Select an appropriate torsion spring for the anticipated range of und-rained shear strength. (Note: Consult the manufacturer's literature for the vane device.) Record the number of the spring used.
2. Make sure that the end of the test sample is squared off.
3. Set the needle on the torque scale at the end of the device to ZERO.
4. Push the vane slowly into the sample, making sure to keep the vane perpendicular to the end of the sample. The vane should penetrate into the sample until the depth mark on the vane shaft is even with the top of the sample.
5. Immediately after reaching the correct vane depth, begin rotating the vane at a rate of 1–1.5 degrees per sec.
6. Watch the needle and read and record the peak (maximum) torque reading from the pointer on top of the apparatus. Also record the spring deflection in degrees.
7. If the post-peak strength is desired, continue rotating the vane until a rotation of 90° is achieved. Read and record this value as the post-peak torque.

8. In order to determine the remolded strength, rotate the vane 10 times and repeat the procedure above to obtain a determination of maximum torque. (Note: ASTM D4648 suggests 5–10 revolutions of the vane; however, the author suggests 10 revolutions as routine practice.)

9. After completing the test, obtain a small soil sample from the test region and determine the Water Content using the convection oven method described in Chapter 1.

Calculations

The undrained shear strength is related to the torque required to rotate the vane. Several assumptions are required relative to the shearing surface, isotropy of the strength, etc. However, the calculations to convert torque to undrained shear strength for a given vane geometry if using a torsion spring type of vane device are straightforward.

$$T = s_u \times K \tag{24.1}$$

where:

T = torque (N-m or lbf-ft.)
s_u = undrained shear strength (Pa or lbf/ft^2)
K = vane constant (m^3 or ft^3)

The vane constant, K, is obtained from:

$$K = \left[\left(\pi D^2 H\right)/2 \times 10^6\right]\left[1 + (D/3H)\right] \text{(SI units)}$$

$$K = \left[\left(\pi D^2 H\right)/3456\right]\left[1 + (D/3H)\right] \text{(in-lbf units)}$$

where:

H = height of vane (mm or in)
D = diameter of vane (mm or in)

For a 12.7 mm × 12.7 mm (0.5 in × 0.5 in) vane: $K = 4.28 \times 10^{-6}$ m^3 (0.0001515 ft^3)

For a 12.7 mm × 25.4 mm (0.5 in × 1.0 in) vane: $K = 7.51 \times 10^{-6}$ m^3 (0.0002651 ft^3)

Since $s_u = T \times k$

$$k = 1/K$$

For a 12.7 mm × 12.7 mm (0.5 in × 0.5 in) vane: $k_1 = 2.34 \times 10^5$ m^{-3} (6600 ft^{-3})

For a 12.7 mm × 25.4 mm (0.5 in × 1.0 in) vane: $k = 1.33 \times 10^5$ m^{-3} (3772 ft^{-3})

The torque, T, is determined from the spring deflection:

$$T = \Delta / B \; (\text{N-m})$$

$$T = \Delta / 12B \; (\text{in.-lbf})$$

where:

Δ = spring deflection in degrees

B = slope of the spring calibration curve in °/N-m (°/lbf-in) provided by manufacturer

Some laboratory vane testers are now available with a direct torque read-out device or have an electronic torque transducer mounted in-line with the vane so that the torque is provided directly. If the vane device provides a direct torque reading, Eq. 24.1 may be used to determine undrained shear strength.

FALL CONE

Introduction

The Fall Cone test was discussed in Chapter 12 for determining the liquid and Plastic Limit. It is perhaps the most attractive of the Index Strength Tests discussed in this chapter. It has been used for over 80 years in Sweden and other countries and has gained increasing popularity in many countries throughout the world. It is simple to use, fast and provides reliable results. A drawing of an early (circa 1915) Fall Cone is shown in Figure 24.14. The test consists of a highly polished stainless steel cone, a drop mechanism and a scale to measure cone penetration. The test may be performed on the end of undisturbed tube samples with the soil still in the tube, or it may be performed on remolded soil by placing the soil in a small cup. A schematic of a modern Fall Cone is shown in Figures 24.15 and 24.16.

Standard

There currently is no ASTM Standard for this test, although a draft standard is in progress.

Equipment

Convection oven capable of maintaining 110°±5°C

Aluminum tins (tare) or ceramic bowls

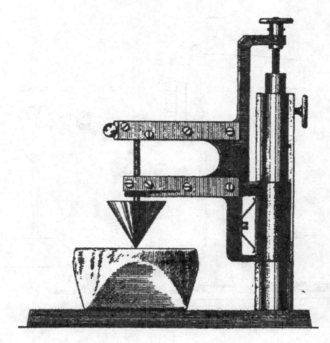

Figure 24.14 Early laboratory Fall Cone (from Skempton 1985).

Electronic scale with a minimum capacity of 500 gm and a resolution of
   0.01 gm
Spatula
Tongs or heat-proof gloves
Desiccator
Fall Cone device with different cones

Procedure

1. Make sure that the end of the test sample is squared off.
2. Select a cone with appropriate mass and angle to suit the consistency
   of the soil. Record the mass and angle.
3. Bring the cone down so that the tip just touches the top of the soil.
4. Set the displacement gauge (dial indicator) to ZERO.
5. Release the cone into the soil. After 5 sec, lock the cone in place and
   record the penetration in mm. If the penetration is less than about 5
   mm or greater than about 10 mm, consider repeating the test with
   either a heavier or lighter cone.
6. If the remolded strength is desired, thoroughly remold the soil in a
   plastic bag, kneading by hand. Fill a small plastic cup with the soil,
   making sure to remove any trapped air and repeat the test. (Note: A

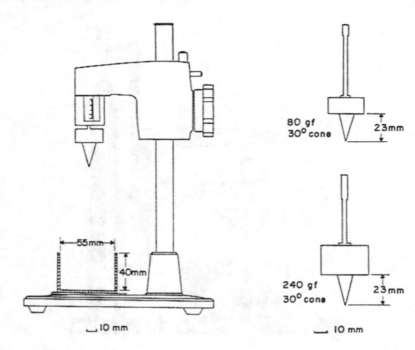

Figure 24.15 Schematic of the Fall Cone test.

Figure 24.16 Fall Cone test and different cones.

lighter cone may be needed for determining the remolded strength depending on the Sensitivity.)

7. After completing the test, obtain a small soil sample from the test region and determine the Water Content using the Convection Oven method described in Chapter 1.

Different styles of Fall Cones are available from different manufacturers; however, the test procedure is the same for all. To perform the test, a cone with the appropriate mass and apex angle is selected to suit the consistency of the soil. Typical cone sizes and the range of strength measured are shown in Table 24.1.

Calculations

The undrained shear strength from a Fall Cone test is determined from:

$$s_u = \left[9.8(M_c)(K)\right]/d^2 \qquad (24.2)$$

where:

s_u = undrained shear strength (kPa)
M_c = mass of the cone (gm)
K = a cone factor which depends on cone apex angle (= 0.8 for 30° cone; = 0.27 for 60° cone)
D = depth of penetration (mm)

Hansbo (1957) and others showed that the results obtained from cones of different mass and cone angle are essentially the same and match the theoretical analysis, provided that cone factors of 0.27 for a 60° cone and 0.8 for a 30° cone are used. While some investigators have suggested some subtle refinements to these cone factors, there have not been any overwhelming refinements that warrant different cone factors.

Table 24.1 Range of Strength Measurement for Different Fall Cones

Cone mass (gm)	Cone apex angle (°)	Range of s_u (kPa)
400	30	10–250
100	30	25–65
80	30	10–200
60	60	0.5–11
10	60	0.08–2

SENSITIVITY

Sensitivity, S_t, is defined as the ratio of undisturbed undrained shear strength to the fully remolded undrained shear strength of the same soil at the same Water Content:

$$S_t = s_{u(und)}/s_{u(rem)} \tag{24.3}$$

The term *Sensitivity* is reserved for the behavior of saturated fine-grained soils. It appears that Terzaghi and Peck (1948) first suggested the term *Sensitivity* and based the definition on laboratory undisturbed and remolded unconfined compression tests. Currently, Sensitivity may be determined using a variety of tests, including the ones described in this chapter, so that there may be some differences observed.

Sensitivity and Liquidity Index

In the past, local correlations between Sensitivity and Liquidity Index (Chapter 12) have been suggested., e.g., Bjerrum 1954. However, more recent compilations of data from the literature for soils from around the world generally show a very poor global relationship between Liquidity Index and Sensitivity (e.g., Shimobe & Spagnoli 2019). There are two primary reasons for this; 1) the determination of both Sensitivity and Liquidity Index have inherent variability because of differences in test methods used; and 2) the process responsible for developing Sensitivity in a fine-grained deposit (e.g., salt leaching) is not the same for all sites, especially if both marine and non-marine sites are included in the same database.

Several laboratory studies have shown that Sensitivity obtained from different test methods is not the same (e.g., Eden & Hamilton 1956; Eden & Kubota 1961; Yong & Tang 1983; Chaney & Richardson 1988; Winters et al. 1990; Abuhajar et al. 2010), and different values are obtained from different test procedures. It appears that this is largely the result of different remolding techniques to obtain the remolded undrained strength. The remolded strength should represent the lowest possible undrained shear strength at a given Water Content. Remolding a sample by rotating a laboratory vane may not produce the same degree as remolding by hand kneading.

Unfortunately, this may also depend on the soil, since some tests may not be appropriate for some soils. For example, it may be difficult to obtain reliable measurements of the remolded strength of highly sensitive clays. Sensitivity may also be influenced by the sample disturbance in an otherwise undisturbed sample. If an undisturbed sample experiences any disturbance during the sampling, transport or handling processes, the measured undrained shear strength will likely be lower than the true undisturbed strength. This will lead to lower values of Sensitivity. When reporting Sensitivity, the test method used should be reported.

For a few marine clay sites in the same general geographic area, there may be a more reasonable correlation between Liquidity Index and Sensitivity; however, this should be checked on important projects where Sensitivity may influence design. Figure 24.17 shows some reported results from Norway, Sweden and Canada relating Sensitivity to Liquidity Index. Although there is a general trend of increasing Sensitivity with increasing Liquidity Index for the Scandinavian clays, there is clearly no universal relationship. Other factors, such as salinity of the pore fluid (discussed in Chapter 11), amorphous content and clay mineralogy, may influence the reliability of any such correlation (e.g., Long et al. 2012).

Table 24.2 gives suggested descriptive terms for use with different ranges of measured Sensitivity. An alternative set of definitions was suggested by Rosenqvist (1953) as given in Table 24.3.

In the mid-1960s, Osterman (1965) suggested that for several sites in Sweden a correlation might be established between Sensitivity and the Water Content ratio (ratio of natural Water Content to Liquid Limit). This eliminates the need to determine Plastic Limit in order to calculate Liquidity Index. Very little work has been conducted in this area since that

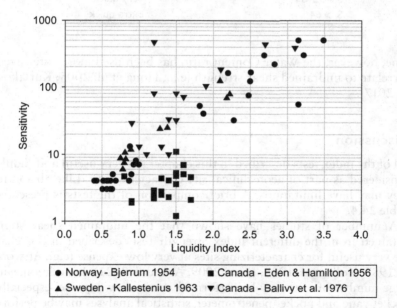

Figure 24.17 Variation in Sensitivity with Liquidity Index for several marine deposits.

Table 24.2 Definitions of Sensitivity (from Skempton & Northey 1952)

Sensitivity	Description
$S_t < 1.0$	Insensitive clays
$S_t = 1-2$	Clays of low Sensitivity
$S_t = 2-4$	Clays of medium Sensitivity
$S_t = 4-8$	Sensitive clays
$S_t = 8-16$	Extra-sensitive clays
$S_t > 16$	Quick-clays

Table 24.3 Definitions of Sensitivity (from Rosenqvist 1953)

Sensitivity	Description
$S_t = 1$	Insensitive
$S_t = 1-2$	Slightly sensitive
$S_t = 2-4$	Medium sensitive
$S_t = 4-8$	Very sensitive
$S_t = 8-16$	Slightly quick
$S_t = 16-32$	Medium quick
$S_t = 32-64$	Very quick
$S_t > 64$	Extra quick

time; however, the Water Content ratio has been used more extensively to correlate to undrained shear strength (e.g., Hong et al. 2005; Kuriakose et al. 2017).

Discussion

All of the index tests described in this chapter are very useful and should be considered as part of geotechnical site characterization. Like all soil tests, they may have limitations. A brief comparison of the tests is presented in Table 24.4.

A number of studies have shown that the undrained shear strength obtained from the different Index Strength Tests described in this chapter are very useful for characterizing sites at very low expense (e.g., Attwooll et al. 1985; Kolk et al. 1988; Leoni 2009; Avunduk 2021). In addition, since a large number of tests may be performed at very low expense, especially by pocket vane and pocket penetrometer, statistical analyses may be performed to evaluate variability in strength characteristics (e.g., Houlsby & Houlsby 2013). Comparisons between the undrained shear strength values obtained by the different tests generally show that the results are very similar (e.g.,

Table 24.4 Comparison of Index Strength Tests

Test	Remarks	Best for	Limitations	Reference
Pocket Penetrometer	Simple calibrated compression spring with direct readings in "equivalent" unconfined compressive strength)	Fine-grained soils up to $q_u = 4.5$ tsf; larger foot very soft clays	Very small test area; maximum reading often attained in very stiff clays	ASTM (in preparation)
Torvane	Simple calibrated torsion spring with direct readings in "equivalent" undrained shear strength)	Fine-grained soils up to $s_u = 2.5$ tsf; larger and smaller vanes for very soft and very hard clays	Not applicable in soils with gravel	ASTM (in preparation)
Fall Cone	Standardized cone with known mass allowed to penetrate surface of soil	Very soft to soft and medium stiff f–g soils	Not applicable for soils with gravel; may need different cones and masses for different soils; cone penetration depth correlated to undrained shear strength	ASTM (in preparation)
Miniature lab vane	Portable hand or electric motor driven miniature vane similar to field vane; maximum torque measured	Very soft to soft f–g soils	Vane will not penetrate stiff f–g soils; not applicable to gravelly soils	ASTM D4648

Goughnour & Sallberg 1964; Sibley & Yamane 1966; Lu & Bryant 1997; Kuomoto & Houlsby 2001; Krizek 2004; Cruz 2009).

As a matter of routine practice, the author recommends that the remolded strength be performed on a few selected samples for all of the tests described in this chapter in order to obtain an indication of Sensitivity. The tests are easy to conduct and only take a small amount of additional time.

REFERENCES

Abuhajar, O., El Naggar, M.H., and Newson, T., 2010. Review of Available Methods for Evaluation of Soil Sensitivity for Seismic Design. *Proceedings of the 54th International Conference on Advances in Geotechnical Earthquake Engineering and oil Dynamics*, Paper 1.32b.

Attwooll, W.J., Fujioka, M.R., and Kittridge, J.C., 1985. Laboratory and In Situ Strength Testing of Marine Soils in Iran and Indonesia. *ASTM STP*, Vol. 883, pp. 440–453.

Avunduk, E., 2021. Possibility of Using Torvane Shear Testing Device for Soil Conditioning Optimization. *Tunnelling and Underground Space Technology*, Vol. 107, p. 12.

Bjerrum, L., 1954. Geotechnical Properties of Norwegian Marine Clays. *Geotechnique*, Vol. 4, pp. 49–69.

Chaney, R.C., and Richardson, G.N., 1988. Measurement of Residual/Remolded Vane Shear Strength of Marine Sediments. *ASTM STP*, Vol. 1014, pp. 166–181.

Cruz, M.P., 2009. Correlation Between Torvane and Unconfined Compression Test in Limos and Loess of Cordoba City. *Proceedings of the 3rd Subamerican Conference on Geotechnical Engineering*.

Eden, W.J., and Hamilton, J.J., 1956. The Use of a Field Vane Apparatus in Sensitive Clay. *ASTM STP*, Vol. 193, pp. 41–53.

Eden, W.J., and Kubota, J.K., 1961. Some Observations on the Measurement of Sensitivity of Clay. *Proceedings of the American Society for Testing and Materials*, Vol. 61, pp. 1239–1249.

Evans, I., and Sherratt, G.C., 1948. A Simple and Convenient Instrument for Measuring the Shearing Resistance of Clay Soils. Army Operations Research Report No. 74, pp. 411–414.

Godskesen, O., 1936. The Spring-Scale Cone, a Pocket Apparatus for Detremining the Firmness of Clay. *Proceedings of the 1st International Conference on Soil Mecahnics and Foundation Engineering*, Vol. 1, pp. 38–40.

Goughnour, R.D., and Sallberg, J.R., 1964. Evaluation of the Laboratory Vane Shear Test. *Highway Research Record*, Vol. 48, pp. 19–29.

Gray, H., 1957. Field Vane Shear Tests of Sensitive Cohesive Soils. *ASCE Transactions*, Vol. 122, pp. 844–857.

Hansbo, S., 1957. A New Approach to Determination of the Shear Strength of Clay by the Fall Cone Test. *Swedish Geotechnical Institute Proceedings*, Vol. 14, p. 47.

Hong, Z., Liu, S., and Negami, T., 2005. Strength Sensitivity of Ariake Clays. *Marine Georesources and Geotechnology*, Vol. 23, pp. 221–233.

Houlsby, N.M., and Houlsby, G.T., 2013. Statistical Fitting of Undrained Shear Strength Data. *Geotechnique*, Vol. 63, No. 14, pp. 1253–1263.

Jakobson, B, 1954. Influence of Sampler Type and Testing Method on Shear Strength of Clay Samples. *Royal Swedish Geotechnical Institute Proceedings*, Vol. 8, p. 59.

Karlsson, R. 1961. Suggested Improvements in the Liquid Limit Test with Reference. *Proceedings of the 5th International Conference on Soil Mechanics and Foundation Engineering*, Vol. 1, pp. 171–184.

Kolk, H.J., Hoope, J.T., and Ims, B.W., 1988. Evaluation of Offshore In Situ Vane Results. *ASTM STP*, Vol. 1014, pp. 339–353.

Koumoto, T., and Houlsby, G.T., 2001. Theory and Practice of the Fall Cone Test. *Geotechnique*, Vol. 51, No. 8, pp. 701–712.

Krizek, R.J., 2004. *Slurries in Geotechnical Engineering. 12th Spencer J. Buchanan Lecture.* Texas A & M University.

Kuriakose, B., Abraham, B.M., Sridharan, A., and Jose, B.T. 2017. Water Content Ratio: An Effective Substitute for Liquidity Index for Prediction of Shear strength of Clays. *Geotechnical and Geological Engineering*, Vol. 35, pp. 1577–1586.

Leoni, A.J., 2009. Characterization of Post Pampeano Clays. *Proceedings of the 17th International Conference on Soil Mechanics and Geotechnical Engineering*, pp. 56–59.

Long, M., Donohue, S., L'Heureux, J.S., Solberg, I.L., Renning, J.S., Limcher, R., O'Conner, P., Sauvi, G., Remoen, M., and Lecomte, I., 2012. Relationship Between Electrical Resistivity and Basic Geotechnical Parameters for Marine Clays. *Canadian Geotechnical Journal*, Vol. 49, pp. 1158–1168.

Lu, T., and Bryant, W.R., 1997. Comparison of Vane Shear and Fall Cone Strengths of Soft Marine Clay. *Marine Georesources and Geotechnology*, Vol. 15, pp. 67–82.

Osterman, J., 1965. Studies on the Properties and Formation of Quick Clays. *Proceedings of the 12th National Conference on Clays and Clay Minerals*, pp. 87–108.

Rosenqvist, I. Th., 1953. Considerations on the Sensitivity of Norwegian Quick-Clays. *Geotechnique*, Vol. 3, pp. 195–200.

Sibley, E.A., and Yamane, G., 1966. A Simple Shear Test for Saturated Cohesive Soils. *ASTM STP*, Vol. 399, pp. 39–47.

Shimobe, S., and Spagnoli, G., 2019. Some Relations Among Fall Cone Penetration, Liquidity Index and Undrained Shear Strength of Clays Considering the Sensitivity Ratio. *Bulletin of Engineering Geology and The Environment*, Vol. 78, pp. 5029–5038.

Skempton, A.W., 1985. A History of Soil Properties: 1717–1927. *Proceedings of the 11th International Conference on Soil Mechanics and Foundation Engineering, Golden Jubilee Volume*, pp. 95–121.

Skempton, A.W., and Northey, R.D., 1952. The Sensitivity of Clays. *Geotechnique*, Vol. 3, pp. 30–53.

Skempton, A.W., and Bishop, A.W., 1950. The Measurement of the Shear Strength of Soils. *Geotechnique*, Vol. 2, No. 2, pp. 90–108.

Terzaghi, K., and Peck, R.B., 1948. *Soil Mechanics in Engineering Practice.* New York: John Wiley Publishers.

Winters, W.J., Dickenson, S.E., and Booth, J.S., 1990. Comparative Effect of Remolding Methods on the Vane Shear Strength of Yellow Sea Sediment. *U.S. Department of Interior U.S. Geological Survey Open File Report 90-213*.

Yong, R.N., and Tang, K.Y., 1983. Soil Remoulding and Sensitivity Measurements. *ASTM Geotechnical Testing Journal*, Vol. 6, No. 2, pp. 73–80.

ADDITIONAL READING

Kallstenius, T., 1963. Studies on Clay Samples Taken with Standard Piston Sampler. *Swedish Geotechnical Institute Proceedings*, Vol. 21, p. 210.

Serota, S., and Jangle, A., 1972. A Direct-Reading Pocket Shear Vane. *Civil Engineering Magazine, ASCE*, Vol. 42, No. 1, January, pp. 73–76.

Sridharan, A., El-Shafei, A., and Miura, N., 2002. Mechanisms Controlling the Undrained Shear Strength of Remolded Ariake Clay. *Marine Georesources and Geotechnology*, Vol. 20, pp. 21–50.

Chapter 25

Thixotropy

INTRODUCTION

Most clays exhibit an increase in remolded undrained shear strength if allowed to rest over time. This behavior is called *Thixotropy* and has been observed in both natural clays and compacted clays. Thixotropy is the increase in undrained shear strength of a saturated clay at constant Water Content with time after remolding. It appears, however, that this behavior is complex and may depend on a number of factors related to soil composition such as mineralogy, clay content, Cation Exchange Capacity, Specific Surface Area and Activity, as well as initial state, which may be described by the Liquidity Index, i.e., relative Water Content with respect to Liquid and Plastic Limits (discussed in Chapter 12). There may also be some influence of pore fluid chemistry on Thixotropy.

Figure 25.1 shows the increase in undrained shear strength measured by Fall Cone of two clay samples of Connecticut Valley varved clay (CVVC) from the same site with time. The behavior is non-linear with a fairly rapid increase in strength at first, and then a slowdown of the rate of strength increase. The results may also be plotted on a semi-log plot as shown in Figure 25.2.

In this case, the results of Fall Cone tests conducted on two different samples from the same site but at different initial Water Content (and Liquidity Index) show a considerable increase in remolded strength with age over a period of 300 days. The Thixotropic Strength Ratio (ratio of aged remolded undrained strength to initial or immediate remolded undrained strength) is shown in Figure 25.3 and is similar for both soils, which is expected since the soils are of the same geology and similar composition.

Early laboratory studies investigating thixotropic behavior of soils have used unconfined compression tests (e.g. Moretto 1948; Berger & Gnaedinger 1949; Jacquet 1990). Laboratory Vane tests (e.g. Skempton & Northey 1952; Goughnour & Sallberg 1964; Hui & Fatt 1971; Shen et al. 2005; Jeanjean 2006; Chan 2015), and Fall Cone tests (e.g., Hansen 1950; Kallstenius 1971; Andersen & Jostad 2002; Lutenegger 2017) have also been used to study thixotropic behavior. The Fall Cone is reliable and easy

DOI: 10.1201/9781003263289-27

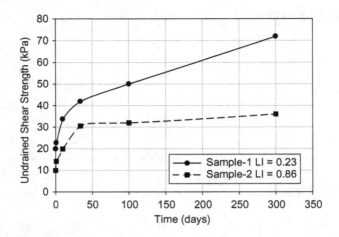

Figure 25.1 Thixotropic behavior of remolded Connecticut Valley varved clay.

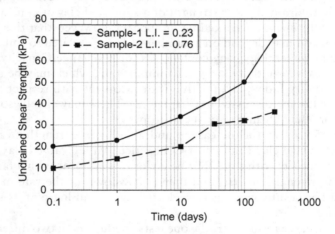

Figure 25.2 Thixotropic behavior of remolded Connecticut Valley varved clay.

to perform and is applicable over a wide range of consistency by using different cones. In this chapter the use of the pocket vane (torvane) and pocket penetrometer are also included as a simple method to determine undrained shear strength as discussed in Chapter 24.

STANDARD

There is no current ASTM Standard for determining thixotropic behavior of clays. The methods described in this chapter use the laboratory Index Strength Tests described in Chapter 24.

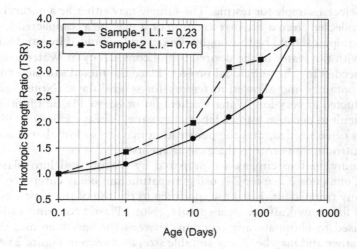

Figure 25.3 Thixotropic Strength Ratio with time.

EQUIPMENT

Convection oven capable of maintaining 110°±5°C
Aluminum tins (tare) or ceramic bowls
Electronic scale with a minimum capacity of 500 gm and a resolution of 0.01 gm
Spatula
Tongs or heat-proof gloves
Desiccator
Plastic specimen containers
Heavy-duty plastic wrap
Heavy-duty aluminum foil

PROCEDURE

The procedure for evaluating thixotropic behavior involves preparing a series of test specimens at constant Water Content and performing laboratory Index Strength Tests.

PREPARATION OF TEST SPECIMENS

Specimens for evaluating the thixotropic behavior of spoils must be prepared from a soil sample that is uniform. Use the following procedure to prepare individual test specimens.

1. Select a sample for testing. The sample may either be a natural sample collected from a site investigation or it may be a compacted sample.
2. Samples should be thoroughly mixed and remolded by hand kneading with tap water to the desired initial consistency or Water Content as needed for the individual project. Place the mixed sample in a sealed container and allowed to temper for several days. (Note: Tests conducted at very low Water Content [at or below Plastic Limit] are difficult to perform using a laboratory vane. A heavy Fall Cone (400 gm) is often needed, and the penetration may be so low that it is easy to introduce an error in the measurements).
3. Individual specimens are prepared by placing soil into plastic Fall Cone cups (Chapter 24) using a spatula and being careful to remove any trapped air. Level the specimen to the top of the plastic cup with a flat draw knife or steel spatula. (Note: Plastic containers should be used to eliminate any reaction between the specimen and the container and may be of any suitable size, as shown in Figure 25.4.)
4. After leveling the top of the specimens, specimens should be double wrapped in plastic wrap and taped with cellophane tape and then double wrapped with aluminum foil. Place the specimens face up in a plastic container and place the lid on the container. Store the container in a constant temperature/humid room near 90% humidity until time for testing.
5. One of the specimens from remolding should be tested immediately after remolding. This specimen can be taken to represent an aging period of 0.0001 days (2 min) in order to plot results on a semi-log scale.
6. At selected times, remove an individual specimen for testing. (Note: Recommended minimum times for aging are 0.1 days [144 min], 1 day, 10 days, and 30 days, requiring 5 individual specimens. Longer aging periods may be used as required for individual projects). Figure 25.4

 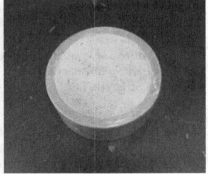

Figure 25.4 Plastic soil cups for aging remolded specimens for Thixotropy testing.

shows a typical plastic soil cup (unfilled and filled) with an inside diameter of 70 mm (2.75 in) and an inside height of 38 mm (1.5 in) for holding remolded soil specimens.

MEASUREMENT OF UNDRAINED SHEAR STRENGTH

1. Select a test method for determining undrained shear strength. One of the Index Strength Tests described in Chapter 24 may be used. Preferably, Fall Cone or laboratory vane tests should be used; however, it is also possible to use pocket vane or pocket penetrometer tests. (Note: The same testing equipment and test procedure should be used to perform all undrained shear strength tests for a given set of specimens from the same soil sample.)
2. At desired aging times, remove specimens from the humid room and perform the undrained shear strength test. After completing the shear strength test, obtain a sample from the middle of the soil cup and determine the Water Content using the convection oven method described in Chapter 1.
3. After performing the final undrained shear strength, prepare a plot of time vs. undrained shear strength as shown in Figure 25.1. Figure 25.5

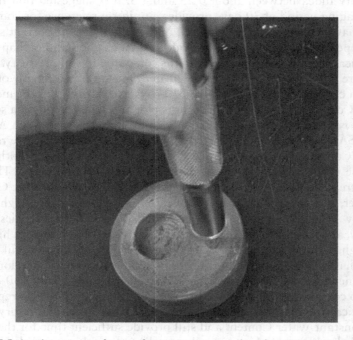

Figure 25.5 Aged specimen after performing torvane and pocket penetrometer tests.

shows an aged soil specimen in a plastic soil cup after performing a torvane and pocket penetrometer test.

CALCULATIONS

Calculations for this test procedure are described for individual laboratory Index Strength Tests and should be followed. Calculate the Thixotropic Strength Ratio (TSR) as:

$$TSR = s_{u30}/s_{u0} \tag{25.1}$$

where:
s_{u30} = undrained shear strength after 30 days
s_{u0} = undrained shear strength at time = 0.001 day

DISCUSSION

The thixotropic behavior of remolded clays and other fine-grained soils is complex and is influenced by both composition and state, although composition appears to be more important than state, at least for soils with a Liquidity Index between about 0.25 and 1.5. It is suggested that the TSR for an aging period of 30 days provides a uniform reference for comparing the behavior of different soils. Based on the results of the current study, it also appears that Activity may provide at least as good and perhaps better an indication of thixotropic behavior as compared to soil plasticity.

Figure 25.6 shows the variation in undrained shear strength of Texas sodium clay for aging periods of 1–310 days as a function of the measured Water Content for each test. The results plot as a linear trend on a semi-log plot as is typical for remolded clays. As indicated, this clay (high Activity; high SSA) shows considerable increase in strength over the entire range of Water Content. Results for this clay indicated that strength continues to increase up to 310 days over the entire range of Water Content. The relative increase in strength appears to be greatest for very high Water Content; however, this may be in part because the initial strength at high L.I. is very low. The development of internal bonds between clay particles occurs quickly, and increase in strength should be expected in a clay with high SSA and high Cation Exchange Capacity (CEC). Beyond 300 days, it is likely that this behavior would show a slowdown in additional strength development.

Lutenegger (2017) suggested that a reference aging period of 30 days be used to compare the thixotropic behavior of different soils. This represents a practical and reasonable time to allow soil to age in the laboratory, maintain constant Water Content and still provide sufficient time for the thixotropic behavior to develop. An understanding of the thixotropic behavior of

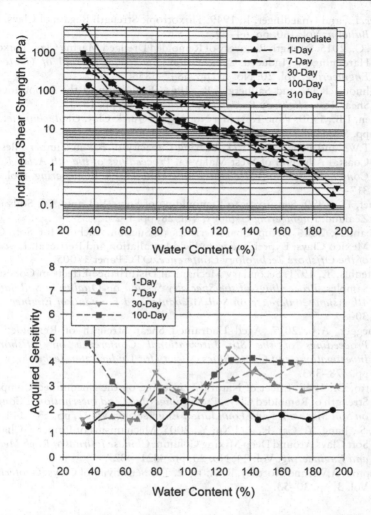

Figure 25.6 Thixotropic behavior of Texas sodium clay over a range of Water Content.

fine-grained soils may be useful in evaluating the regain in strength around driven piles in clays, installation of onshore and offshore anchors and other problems involving disturbance or remolding of clay and subsequent behavior in undrained loading.

REFERENCES

Andersen, K., and Jostad, H., 2002. Shear Strength Along Outside Wall of Suction Anchors in Clay After Installation. *Proceedings of the 12th International Offshore and Polar Engineering Conference*, pp. 785–794.

Berger, L., and Gnaedinger, J., 1949. Thixotropic Strength Regain of Clays. *ASTM Bulletin*, Vol. 160, pp. 64–69.

Chan, C., 2015. Strength Recovery of Remolded Dredged Marine Clay: Thixotropic Hardening vs. Induced Cementation. *Electronic Journal of Geotechnical Engineering*, Vol. 20, No. 13, pp. 5847–5858.

Goughnour, R.D., and Sallberg, J.R., 1964. Evaluation of the Laboratory Vane Shear Test. *Highway Research Record*, Vol. 48, pp. 19–33.

Hansen, J.B., 1950. Vane Tests in a Norwegian Quick-Clay. *Geotechnique*, Vol. 2, pp. 58–63.

Hui, T.W., and Fatt, C.S., 1971. Bearing Capacity of Bakau Timber Piles in the Coastal Alluvium of West Malaysia. *Proceedings of the 4th Asian Regional Conference on Soil Mechanics and Foundation Engineering*, Vol. 1, pp. 317–321.

Jacquet, D., 1990. Sensitivity to Remoulding of Some Volcanic Ash Soils in New Zealand. *Engineering Geology*, Vol. 28, pp. 1–25.

Jeanjean, P., 2006. Setup Characteristics of Suction Anchors for Soft Gulf of Mexico Clays: Experience from Field Installation and Retrieval. *Proceedings of the Offshore Technology Conference*, OTC Paper 18005.

Kallstenius, T., 1971. Secondary Mechanical Disturbance Effects in Cohesive Soil Samples. *Proceedings of the Specialty Session on Quality in Soil Sampling. 4th Asian Conference on Soil Mechanics and Foundation Engineering*, pp. 30–39.

Lutenegger, A.J., 2017. Aged Undrained Shear Strength of Remolded Clays. *Proceedings of the 8th International Conference on Offshore Site Investigations and Geotechnics, Society for Underwater Technology*, Vol. 1, pp. 378–383.

Moretto, O., 1948. Effect of Natural Hardening on the Unconfined Compression Strength of Remolded Clays. *Proceedings of the 2nd International Conference on Soil Mechanics and Foundation Engineering*, Vol. 2, pp. 137–144.

Shen, S., Jiang, Y., Cai, F., and Xu, Y., 2005. Mechanisms of Property Changes of Soft Clays Around Deep Mixing Column. *Chinese Journal of Rock Mechanics and Engineering*, Vol. 24, No. 23, pp. 4321–4327.

Skempton, A.W., and Northey, R.D., 1952. The Sensitivity of Clays. *Geotechnique*, Vol. 3, pp. 30–53.

ADDITIONAL READING

Kamil, A.S., and Aljorany, A.N., 2019. Thixotropic Hardening of Fao Clay. *Journal of Engineering*, Vol. 25, No. 5, pp. 68–78.

Mitchell, J.K., 1961. Fundamental Aspects of Thixotropy in Soils. *Transactions of the ASCE*, Vol. 126, No. 1, pp. 1586–1620.

Ren, Y., Yang, S., Anderson, K.H., Yang, Q., and Wang, Y., 2021. Thixotropy of Soft Clay: A Review. *Engineering Geology*, Vol. 287, p. 9.

Ruge, J.C., Molina-Gomez, F., and Rojas, J.P., 2019. Thixotropic Behavior Study of Clayey Soils from the Lacustrine Deposits of Bogota High Plateau. *Journal of Physics: Conference Series*, Vol. 1386, p. 7.

Seed, H.B., and Chan, C.K., 1957. Thixotropic Characteristics of Compacted Clays. *Journal of the Soil Mechanics and Foundations Division*, Vol. 83, No. 4, pp. 1427-1–1427-35.

Tanaka, H., and Seng, S., 2013. Hardening Process of Clayey Soils with High Water Content Due to Thixotropy Effect. *Proceedings of the 18th International Conference on Soil Mechanics and Geotechnical Engineering*, pp. 433–436.

Tang, B., Zhou, B., Xie, L., Yin, J., Zhao, S., and Wang, Z., 2021. Strength Recovery Model of Clay During Thixotropy. *Advances in Civil Engineering*, Vol. 2021, Article 8825107, p. 11.

Yang, S., and Andersen, K.H., 2016. Thixotropy of Marine Clays. *ASTM Geotechnical Testing Journal*, Vol. 39, No. 2, p. 9.

Zhang, X.W., Kong, L.W., Yang, A.W., and Sayem, H.M., 2017. Thixotropy Mechanism of Clay: A Microstructural Investigation. *Soils and Foundations*, Vol. 57, pp. 23–35.

Zreik, D., Germaine, J., and Ladd, C., 1998. Effect of Aging and Stress History on the Undrained Shear Strength of Ultra-Weak Cohesive Soils. *Soils and Foundations*, Vol. 38, No. 4, pp. 31–39.

Index

Activity, 158
Atterberg Limits, 133, 134, 155

British Soil Classification System, 157

Calcite, 80, 81
Carbonate Content, 71, 81
Casagrande Plasticity Chart, 156
Cation Exchange Activity, 94
Cation Exchange Capacity, 91, 95, 106
Chittick Apparatus, 76
Clay, 42, 52
Coefficient of Curvature, 45, 46
COLE, 187
Consistency Index, 154
Convection oven, 3
Corrosivity, 264

Density, 23, 30
Differential Free-Swell Index, 203
Dispersion
    determination by crumb test, 218
    determination by double
        hydrometer, 210
    determination by pinhole test, 212
Dolomote, 76, 80, 81

Electrical resistivity
    determination by four electrode
        method, 258
    determination by two electrode
        method, 260
Eley Volumeter, 30
European Soil Classification
        System, 157

Fall Cone, 140, 284, 286
Free-Swell Index, 198, 206
Free-Swell Ratio, 203

Grain-size distribution curve, 46

Hydrogen peroxide, 64
Hydrometer analysis, 47
Hygroscopic Water Content, 231, 234

Initial Consumption of Lime, 221, 224

Kaolinite, 38, 95, 201, 205

Laboratory miniature vane test,
        280, 282
Lime Fixation Point, 227, 229
Linear Shrinkage, 181, 187
Liquidity Index, 154, 288
Liquid Limit
    determination by Casagrande drop
        cup, 134, 145
    determination by Fall Cone,
        139, 145
Liquid Limit Ratio, 66
Loss-on-ignition test, 61

Methylene Blue Index, 113, 114
Methylene Blue Test, 111
Microwave oven, 8
Modified Free-Swell Index, 200
Montmorillonite, 38, 95, 201, 205

One-Point Liquid Limit Test
    Casagrande drop cup, 153
    Fall Cone, 154
Organic Content, 59, 67
Organic soils, 59

Peat, 60
pH, 85
Phase relationships, 19
Pipette analysis, 51

Plastic Limit, 147, 150
    determination by Fall Cone, 150
    determination by rolling
        method, 147
Plasticity Index, 154
Plasticity Ratio, 158
Pocket penetrometer, 271, 274

Rapid Carbonate Analyzer, 72
Refractometer, 118, 121
Remolded undrained shear
        strength, 152

Salinity, 117, 122
Saturation, 22
Sensitivity, 117, 126, 288, 290
Shrinkage Index, 154
Shrinkage Limit
    determination by direct
        method, 175
    determination by mercury
        method, 166
    determination by wax method, 171
Shrinkage Ratio, 171
Shrink Test, 191, 195
Sieve analysis, 42
Soil classification
    British Soil Classification System, 157
    European Soil Classification
        System, 157

Unified Soil Classification System,
        155, 156
Soil press, 119
Specific Gravity, 33, 38
Specific Surface Activity, 105
Specific Surface Area, 97, 104,
        106, 113
Speedy Moisture Meter, 11
Suction, 237

Ternary diagram, 57
Thermal Conductivity, 245, 250
Thermal needle, 245, 246
Thixotropic Strength Ratio, 295
Thixotropy, 295
Torvane, 276

Uniformity Coefficient, 44, 46
Unit Weight, 22, 24, 30

Voids Ratio, 21

Water adsorption, 231
Water Content
    determination by convection
        oven, 3
    determination by gas carbide
        tester, 11
    determination by microwave
        oven, 8

Printed in the United States
by Baker & Taylor Publisher Services